Infrastructure
Management

Infrastructure Management

Integrating Design, Construction, Maintenance, Rehabilitation, and Renovation

W. Ronald Hudson
Ralph Haas
Waheed Uddin

McGraw-Hill

New York   San Francisco   Washington, D.C.   Auckland   Bogotá
Caracas   Lisbon   London   Madrid   Mexico City   Milan
Montreal   New Delhi   San Juan   Singapore
Sydney   Tokyo   Toronto

Library of Congress Cataloging-in-Publication Data

Hudson, W. Ronald.
    Infrastructure management : integrating design, construction,
maintenance, rehabilitation, and renovation / W. Ronald Hudson,
Ralph Haas, Waheed Uddin.
      p.   cm.
    Includes index.
    ISBN 0-07-030895-0
    1. Engineering—Management. 2. Infrastructure (Economics)—
Management. 3. Life cycle costing. 4. Engineering economy.
I. Haas, R. C. G. (Ralph C. G.) II. Uddin, Waheed. III. Title.
TA190.H82   1997
363'.068—dc21   97-16887
                                                       CIP

McGraw-Hill

A Division of The McGraw·Hill Companies

  2 3 4 5 6 7 8 9 0 DOC/DOC 9 0 2 1 0 9 8

ISBN 0-07-030895-0

*The sponsoring editor for this book was Larry Hager, the editing supervi-
sor was Bernard Onken, and the production supervisor was Clare Stanley.
It was set in Century Schoolbook by Victoria Khavkina of McGraw-Hill's
Professional Book Group composition unit.*

Printed and bound by R. R. Donnelley & Sons Company.

McGraw-Hill books are available at special quantity discounts to use
as premiums and sales promotions, or for use in corporate training programs. For more information, please write to the Director of Special
Sales, McGraw-Hill, 11 West 19th Street, New York, NY 10011. Or
contact your local bookstore.

This book is printed on recycled, acid-free paper containing
a minimum of 50% recycled, de-inked fiber.

Contents

Preface

A great deal of thought went into titling this book. We considered *Infrastructure Management, Asset Management,* and *Facilities Management* because all of these terms are occasionally used to describe the kinds of activities that are covered in this book. It should be noted that these terms are similar and are often used interchangeably to describe the process of life-cyle cost integration. That is, the process of integrating design, construction, maintenance, rehabilitation, and renovation to maximize benefits to the user and minimize total cost to the owners and users.

There are, however, some subtle differences among the terms, which led to the final choice for this book. The term "asset management" has arisen primarily from use in the private sector. Private corporations in the profit-making business try to manage their "assets," which include physical infrastructure as well as capital, personnel, technology, and technology ideas.

"Facility management" is most often used with a physical entity broadened to include operations and furnishings. In fact, the term facility management is generally meant to describe operation and use of a facility, such as a hotel, including both the infrastructure or building itself as well as furniture, ballrooms, banquet facilities, dishes, linen, and so forth. The activity therefore often includes scheduling, setting up furniture in a ballroom for a banquet, and similar operations.

The term "infrastrucure management" has been coined to generalize the concepts of pavement management, bridge management, and building management and has most often been applied to public civil engineering infrastructure, such as water, waste water, bridges, airports, parks, pavements, and the like.

Much of the technology used in the three areas described is the same, that is, good data inputs, economic models and analysis, benefit cost studies, good maintenance, and rehabilitation are all needed to provide adequate protection of the investment. Given the similarities, but keeping in mind the subtle differences of these three terminologies, the term Infrastructure Management Systems (IMS) was selected for use. We felt that IMS was more descriptive of the process that covers public infrastructure assets and facilities than the other two terms, and we will use this term throughout this book, *Infrastructure Management.* The reader should not be confused by others who may choose to use the word *assets* or *facilities,* they are not "new and different." The concepts presented herein are applicable no matter which choice of wording you select.

This book is intended to provide an overview for those interested in all aspects of civil engineering infrastructure, especially those interested in learning how to procure and preserve it more efficiently and economically. The book can be used by students or practicing engineers to enlighten themselves about the broader aspects of IMS. Each section of the book stands alone, and the book can be read or studied in parts, by those active in one part of the process (i.e., design) who want to learn more about another part of the process, such as maintenance and rehabilitation. Someone in a peripheral area of infrastructure, such as a small town or an airport authority not now using infrastructure management in their agency, can use the book to excite their interest in new applications and developments in their workplace.

Perhaps most importantly, the authors hope that this book will provide a compilation of course materials and an impetus to the development of courses and a change in university curricula to specifically teach engineering students the real need to integrate maintenance and rehabilitation into their future activities. Current curricula too often stress primarily design and construction when, in fact, most engineering work in the next two decades will involve a large component of preservation, maintenance, and rehabilitation, as well as improved use of infrastructure facilities.

This book is dedicated to our respective wives, Martha Hudson, Rose Haas, and Rukhsana (Juhee) Uddin, and to our many colleagues known and unknown who have for the past 30 years helped us to advance the technology and use of better management concepts in all phases of civil engineering infrastructure. Thanks are due to Dr. Zhanmin Zhang, Omar Uddin, and Usman Uddin for their contributions in the preparation of the manuscript. The manuscript was carefully typed by Julie Wickham, Jan Zeybel, and Loretta McFadden, and reviewed by Betty Brown, who are due special thanks for their perseverance.

We are specially indebted to Dr. Zhanmin Zhang, who prepared Chapter 17. He as done pioneering work in developing working IMS computer programs. Dr. Zhang is currently a Research Engineer at the University of Texas in Austin.

The roots of the IMS process lie in pavement management, for which we make no apology because none is necessary. There is proven application of the concepts already in the areas of highways, bridges, water, and waste-water facilities. Only the imagination and the ability of you, the reader, limit the future use of this book and the concepts presented.

W. Ronald Hudson

Ralph Haas

Waheed Uddin

The Challenge of Managing Infrastructure

The Big Picture

1.1 Infrastructure and Society

The success and progress of human society depends on physical infrastructure for distributing resources and essential services to the public. The quality and efficiency of this infrastructure affects the quality of life, the health of the social system, and the continuity of economic and business activity. A nation's economic strength is reflected in its infrastructure assets. Many examples can be cited from history. The Romans built a strong empire by constructing all-weather roads and viaducts throughout Europe, North Africa, and the Middle East to move people, goods, and water. In the colonial era of the 16th to 19th centuries, European nations emerged as strong shipbuilders and explorers. This was followed by the products of the industrial revolution, particularly in the use of steam engines for ships and railroad transport.

In the United States and other regions of the world, historical development of the economic and social systems closely parallels phases of infrastructure development. Demands on infrastructure and related services increase as people expect a higher standard of life and public services. But, more importantly, good infrastructure facilitates a higher quality of life.

1.1.1 Road infrastructure and economic development

Clarification of the relationship between infrastructure and economic development is provided by Queiroz in his World Bank study [Queiroz 92]. It shows a very strong association between economic development, in terms of per capita gross national product (GNP), and road infrastructure. Gross national product is the measure of a nation's total

3

market value of the goods and services that are produced annually. GNP per capita is a country's gross national product divided by its population. Road infrastructure can be characterized by spatial density, which is a country's road length per land area, and road density, which is the length of the road network per capita. Road transport is important to economic activity, especially in developing countries, where it plays an essential role in marketing agricultural products and providing access to health, education, and other services. A good road system gives a country a competitive edge in moving goods efficiently and economically.

Considering the United States as illustrative of a highly developed country, there is a vast amount of historical data available on the road network and the economy [FHWA 91, Abstract 91]. A time-series analysis of this data from 1950 to 1988 shows a significant positive relationship between per capita GNP ($PGNP$, in US\$1,000 per inhabitant, using 1982 constant dollars) and density of paved roads (length of paved roads, LPR, in km per 1,000 inhabitants):

$$PGNP = -3.39 + 1.24(LPR) \tag{1.1}$$

A similar type of analysis for 98 developing countries was carried out by Queiroz and Gautam, with the resulting equation:

$$PGNP = 1.39\,(LPR) \tag{1.2}$$

Figure 1.1 also shows this relationship and the associated statistics. To make a 1988 constant dollar comparison with the developing countries, the time-series equations from the United States data were converted using the GNP implicit price deflator between 1982 and 1988 [Queiroz 92]:

$$PGNP_{88} = -4.1 + 1.50\,(LPR) \tag{1.3}$$

A similar analysis of 1950–1988 analysis of Canadian data [Queiroz 94] yielded the following equation:

$$PGNP_{88} = 0.86 + 1.33\,(LPR) \tag{1.4}$$

This equation is also plotted on Figure 1.1 (i.e., a comparison is provided of eqs. 1.2, 1.3, and 1.4, according to their inference space). There is relatively good consistency between the two equations for 98 countries and Canada. The United States has about 13 percent greater road density for any given $PGNP$ value.

An interesting time lag analysis shows that the highest correlation exists for $PGNP$ of a given year compared with LPR four years earlier (Figure 1.2). It suggests that a paved road investment today will

Figure 1.1 Comparison of data from US, Canada, and 98 developing countries. [after Queiroz 94].

Figure 1.2 Correlation between real per capita GNP and lagged paved road density in the US. [after Queiroz 92].

result in an increased GNP, about four years later. This four-year time lag is in broad agreement with the "half a decade" lag period observed by Aschauer [Aschauer 89].

A second comparison analysis between the supply and condition of paved road networks in 98 developing and developed countries is

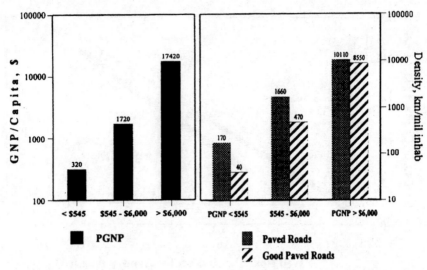

Figure 1.3 Average road density in low-, middle-, and high-income economies. [after Queiroz 92].

shown in Figure 1.3 [Queiroz 92]. The countries are grouped as defined by the World Bank [World 90]:

1. Low-income economies are those with a *PGNP* of $545 or less in 1988 (42 countries).

2. Middle-income economies are those with a *PGNP* of more than $545 but less than $6,000 in 1988 (43 countries).

3. High-income economies are those with a *PGNP* of $6,000 or more in 1988 (13 countries).

Figure 1.3 shows that the supply of good paved road infrastructure in high-income economies is dramatically higher than that in middle- and low-income economies.

A decline in maintenance and capital improvement investments may significantly affect the condition and adequacy of the infrastructure. This has been examined in detail for 23 states by the University of Colorado at Boulder in *Hard Choices: A Report on the Increasing Gap between America's Infrastructure Needs and Our Ability to Pay for Them* [Colorado 84]. The report shows that state and local infrastructure outlays have declined from 2.2 percent of GNP in 1961 to 1.9 percent in 1982 and have reduced across all regions of the US. For the US as a whole, US$450 billion (in 1982 dollars) was the estimated future infrastructure funding gap between available and needed outlays. The most dominant need was identified for highway and bridge infrastructure.

1.1.2   Air transport and economic development

The air-transport industry and airport infrastructure in the United States has evolved in parallel with the nation's economic, industrial, and social development. With over 16,000 public airports [DOT 94], the US has more than the rest of the world combined. The world airport infrastructure and air-transportation system have also continually grown in recent years to meet the air-transport demands of passengers and freight. Between 1979 and 1988, world air traffic in terms of passenger kilometers grew annually at the rate of 5.5 percent and the number of freight kilometers grew at 7.5 percent annually, as compared to the 2.5 percent annual growth of GNP and 4 percent growth of international trade in the OECD (Organization for Economic Cooperation and Development) countries [Veldhuis 92]. These figures indicate the trend of increased air-transport share in the global economy.

1.1.3   Urban infrastructure and economic growth

Business communities, commercial enterprises and industries know the benefits of a well-maintained and well-served infrastructure system. According to *Fortune* magazine [Perry 89], in a 1988 Lou Harris survey of 250 companies conducted for Cushman & Wakefield, a real-estate firm, most respondents rated "easy access to domestic markets, customers, or clients" as the most important factor in choosing an office location. The survey ranked Atlanta, which boasts an impressive array of freeways, mass transit, and air-transportation facilities, as the No. 1 city in which to put a business. Says George Berry, Georgia Commissioner of Industry and Trade, "Transportation services are to Atlanta what gambling is to Las Vegas."

1.2   Infrastructure Definition

Infrastructure has been defined in many ways; for example, "those physical facilities that are sometimes called public works" [Grigg 88]. Public works have been defined by the American Public Works Association (APWA) as follows [Stone 74] :

> Public works are the physical structures and facilities that are developed or acquired by the public agencies to house governmental functions and provide water, power, waste disposal, transportation, and similar services to facilitate the achievement of common social and economic objectives.

The National Science Foundation report on civil infrastructure systems states [NSF 94]:

A civilization's rise and fall is linked to its ability to feed and shelter its people and defend itself. These capabilities depend on infrastructure—the underlying, often hidden foundation of a society's wealth and quality of life. A society that neglects its infrastructure loses the ability to transport people and food, provide clean air and water, control disease, and conduct commerce.

The following infrastructure definition is given by the Associated General Contractors of America (AGCA), for all long-lived assets owned by local, state, and federal governments and utilities owned by businesses [Kwiatkowski 86]:

> The nation's infrastructure is its system of public facilities, both publicly and privately funded, which provide for the delivery of essential services and a sustained standard of living. This interdependent, yet self-contained, set of structures provides for mobility, shelter, services, and utilities. It is the nation's highways, bridges, railroads, and mass transit systems. It is our sewers, sewage, sewage treatment plants, water supply systems, and reservoirs. It is our dams, locks, waterways, and ports. It is our electric, gas, and power-producing plants. It is our court houses, jails, fire houses, police stations, schools, post offices, and government buildings. America's infrastructure is the base upon which society rests. Its condition affects our life styles and security and each is threatened by its unanswered decay [AGCA 82].

The government accounting statements consider infrastructure to be the fixed assets of value to the government which include roads, bridges, curbs and gutters, streets and sidewalks, drainage systems, lighting systems, and similar assets that are immovable and of value only to the governmental unit. [Kwiatkowski 86]. Such a definition, while useful for accounting, ignores the value of the asset to the citizen.

In this book, "infrastructure" refers to all these combined facilities that provide essential public services of transportation, utilities (water, gas, electric), energy, telecommunications, waste disposal, park lands, sports, and recreational and housing. Infrastructure also provides the physical systems used to provide other services to the public through economic and social actions. These infrastructure facilities and services are provided by both public agencies and private enterprises.

1.3 Historical Overview of Infrastructure Development

The trend toward major infrastructure development in the industrial nations, especially in road and bridge construction, began in the 18th century. In the 19th century, major infrastructure development paralleled economic development. Street construction technology was improved by using asphalt and concrete to facilitate all-weather automobile traffic. Most modern roadways have been constructed since 1900. Prior to that, railroads were the primary mode of long-distance surface transportation during the 19th century and early 20th century. Philadelphia was the first city to initiate the construction of a large-scale municipal water supply system in 1798, followed by New York and Boston by the mid-19th century [Grigg 88].

A chronological summary listing of infrastructure milestones in the present century is presented below [Infra 92, Horonjeff 94].

1900–1909. Construction of sanitary and ship canal in Michigan, initiation of the Panama Canal project; construction of 320-km (200-mi) long Owens River aqueduct to bring water to Los Angeles; introduction of gasoline and diesel power to roller and excavator; Wright Brothers' first airplane flight; opening of New York City's subway system; first concrete road paved in Wayne County, Michigan; 1906 earthquake that ravaged San Francisco.

1910–1919. Roosevelt Dam constructed near Phoenix, Arizona; construction of the first transcontinental highway (Lincoln Highway U.S. 30); introduction of motor grader; start of World War I in 1914 and conclusion in 1918; first U.S. federal-aid highway program legislation in 1916, with 50 percent federal participation; gasoline tax to pay for roads; postconstruction boom of roads, rural highway network grows to 4 million km (2.5 million mi) with 10 percent surfaced, including 16,800 km (10,500 mi) of asphalt, 3680 km (2300 mi) of concrete, and 2560 km (1600 mi) brick roads; first air-mail service between Washington, D.C. and New York in 1918.

1920–1929. Use of pneumatic rubber tires; 9 million autos and trucks damaging the unbound roads; introduction of earth scrapers, mobile cranes and tractors; assigning national road-numbering system (even numbers to east-west routes and odd numbers to north-south; main transcontinental routes numbered in multiples of 10); completion of the 9.8-km (6.1-mi) long Moffectrail tunnel near Denver and the Holland tunnel under the Hudson river between Manhattan and Jersey City; opening of intercontinental air travel; completion of Newark's international airport; 1928 Air Commerce Act; stock market crash in 1929 and start of the Great Depression.

1930–1939. President Roosevelt's National Recovery Act (NRA), which included a large program of public works construction and $400 million to the states for highway construction; completion of the Detroit-Windsor tunnel, New York City's George Washington Bridge, Hoover Dam, Bonneville Dam and hydroelectric power station in

Oregon, San Francisco's Trans Bay Bridge; Golden Gate suspension bridges; the Blue Ridge Parkway; establishment of the Tennessee Valley Authority (TVA); 1938 Civil Aeronautics Act defining federal government role in aviation; Germany's invasion of Poland and start of World War II.

1940–1949. Creation of the Pentagon; construction of military air fields; construction of the 2270 km (1420 mi) Alaska Highway, the Pennsylvania Turnpike, the Grand Coulee Dam (concrete construction) and hydro-electric power station; nuclear research leading to the atomic bomb and later to nuclear power generation; end of the war in 1945; slip-form concrete paver introduced; Federal-Aid Highway Act of 1944 authorizing more than $1.6 billion over three years and the designation of a National System of Interstate Highways; Federal Airport Act of 1946; postwar construction of the Maine Turnpike and conversion of 500 military airports to civilian use.

1950–1959. Large-scale toll-road construction started with the 189-km (118-mi) $225 million New Jersey Turnpike; construction of the first prestressed concrete bridge (the Walnut Lane Bridge in Philadelphia) and the first segmented concrete bridge (the Griffith Road Bridge in New York State); start of hauling of truck trailers on rail flat-bed cars ; construction of the nation's first commercial nuclear power plant in Pennsylvania; the first transcontinental telephone cable in operation; Federal Aid Highway Act of 1956 authorizing 65,600 km (41,000 mi) of National System of Interstate and Defense Highways at an estimated $41 billion and establishment of the Highway Trust Fund; construction and testing of the $27 million AASHO Road Test to produce data for improved highway design; Federal Aviation Act of 1958 establishing the Federal Aviation Agency.

1961–1969. Space Age initiated with the first American in orbit in 1962, the first moonwalk in 1969, and construction of National Aeronautical Space Administration (NASA) facilities; construction of the 28-km (17.6-mi) long, $200 million Chesapeake Bay Bridge-Tunnel; construction of the Oroville Dam (earthen fill dam) in Northern California; start of construction on 26 nuclear power plants; passage by Congress of the Water Quality Act in 1965, the National Environmental Protection Act in 1969, and creation of a cabinet-level U.S. Department of Transportation (DOT) in 1967, which included the Federal Highway Administration, the Urban Mass Transportation Administration, the Federal Aviation Administration, the Federal Railroad Administration and Maintenance Administration; collapse of the 39-year-old Silver Bridge over the Ohio River and, as a result, development of national bridge-inspection standards and conception of a bridge-replacement program.

1970–1979. Authorization by Congress of $250 million to start the bridge-replacement program; establishment of the Occupational Safety and Health Administration (OSHA), U.S. Environmental Protection Agency (EPA); Airport and Airway Development Act of 1970, Airline Deregulation Act of 1978, and Aviation Safety and Noise Abatement Act of 1979; passages by Congress of the Clean Air Act; Federal Water Pollution Act authorizing $18 billion for construction of waste-water treatment plants; authorization by Congress of the 1280-km (800-mi) long Alaskan Oil pipeline and tanker terminal; construction of JFK Causeway in Corpus Christi, a cable-stayed bridge in Alaska and the first cast-in-place segmented Pine Valley bridge near San Diego; opening of the Bay Area Rapid Transit (BART) System in San Francisco; Federal Aid Highway Act permitting limited use of federal funds for urban mass-transit projects; first trial of asphalt recycling for road construction; first Earth Day in 1970.

1980–1989. Rebuilding highways without shutting down traffic; Congress expands the 4R program with the federal match of 90 percent; the Surface Transportation Assistance Act of 1982 raising the fuel tax on gasoline from 4 to 9 cents per gallon, with 1 cent earmarked for mass transit; 72,000-km (45,000-mi) long interstate system near completion; Airport and Airway Improvement Act of 1982; construction of the first roller-compacted Willow Creek Dam in Oregon; earthquakes creating major damage in Mexico and California.

1990–1992. Updating of Clean Air Act in 1990 with strict regulation on reduced vehicle emissions; construction decline due to recession; Airway Safety and Capacity Expansion Act of 1990; the 1991 Intermodal Surface Transportation Efficiency Act (ISTEA) authorizing $155 billion over six years with emphasis on maintenance, service life analysis, and management systems; 1992 Chicago flood and Hurricane Andrew in South Florida; rebuilding infrastructure in Kuwait after the 1991 Gulf War, attracting U.S. and other international contractors; other significant global changes, including the collapse of the Berlin Wall, opening of East European countries to the West, breakup of Soviet Union into independent republics and end of Cold War.

Post-1992. Recession in the US and world economies; increased emphasis to reduce federal spending, including the administrative cost in US DOT budget and Amtrak budget; continuation of highway rehabilitation projects and establishment of a high-quality national highway system (NHS) consisting of nearly 256,000 km (160,000 mi); opening of the new Denver International Airport; construction of Olympic facilities in Atlanta; the 1994 ice-storm damage to electric-power stations and water-supply systems in the Midsouth region; enormous infrastructure damages during the 1994 Los Angeles earthquake in the United States, and the 1995 Kobe earthquake in Japan.

1.4 Infrastructure Assets

1.4.1 Categories of infrastructure facilities

Physical facilities and related services can be categorized in seven groups based on their primary functions and services, as shown in Table 1.1.

TABLE 1.1   Categories of infrastructure assets.

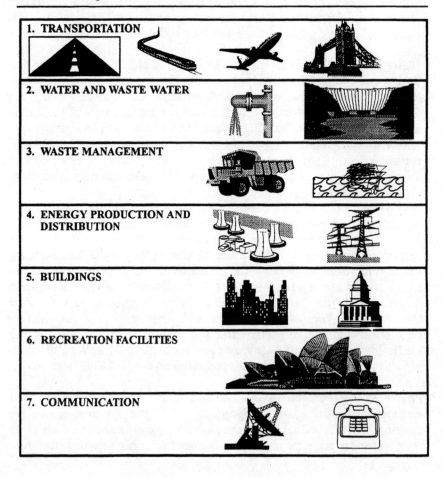

In the United States, most of the roads, airports, urban mass-transit infrastructure facilities, water and wastewater facilities, and waste-management facilities are built and maintained by public agencies. Both public agencies and private enterprises are involved in energy production and distribution infrastructure, buildings, and recreational facilities. Railroad facilities and communications infrastructure are historically built and managed by the private sector.

The following subgroups of infrastructure assets include most of the publicly and privately owned and managed facilities used to provide the essential services to support and sustain the civilian life in a nation.

1.4.1.1 Transportation

- Ground transportation (roads, bridges, tunnels, railroads)
- Air transportation (airports, heliports, ground facilities, air-traffic control)
- Waterways and ports (shipping channels, main terminals, dry docks, ports)
- Intermodal facilities (rail/airport terminals, truck/rail/port terminals)
- Mass transit (subways, bus transit, light rail, monorails, platforms/stations)

1.4.1.2 Water and waste water

- Water supply (pumping stations, treatment plants, main water lines, wells, mechanical/ electric equipment)
- Structures (dams, diversion, structures, tunnels, aqueducts)
- Agricultural water distribution (canals, rivers, weir, gates, dikes)
- Sewer (main sewer lines, septic tanks, treatment plants, stormwater drains)

1.4.1.3 Waste management (landfills, treatment plants, recycling facilities)

- Solid waste
- Hazardous waste
- Nuclear waste

1.4.1.4 Energy production and distribution

- Electric power production (hydro-electric power stations; gas-, oil-, and coal-fueled power generation)

- Electric power distribution (high-voltage power transmission lines, substations, distribution systems, energy-control center, service and maintenance facilities)
- Gas pipeline (gas production, pipeline, computer stations and control centers, storage tanks, service and maintenance facilities)
- Petroleum/oil production (pumping stations, oil/gas separation plants, roads)
- Petroleum/oil distribution (marine and ground tanker terminals, pipelines, pumping stations, maintenance facilities, storage tanks)
- Nuclear power stations (nuclear reactors, power-generation stations, nuclear waste disposal facilities, emergency equipment and facilities)

1.4.1.5 Buildings

- Tall buildings—residential/commercial (structures, utilities, swimming pools, security, ground access, parking)
- Public buildings (schools, hospitals, government offices, police stations, fire stations, postal offices, parking structures)
- Multipurpose complexes (coliseums, amphitheaters, convention centers, religious congregation buildings)
- Sports complexes (indoor, stadiums, golf courses)
- Movie theaters (indoor, drive-in)
- Housing facilities (public, private)
- Manufacturing/warehouse facilities
- Hotels/commercial properties (hotels, inns, time-sharing units, clubs, malls)

1.4.1.6 Recreational facilities

- Parks and playgrounds (roads, parking areas, recreational facilities, office buildings, rest rooms, ornamental fountains, swimming pools, picnic areas)
- Lake and water sports (roads, parking areas, picnic areas, marinas)
- Theme parks/casinos (access roads, buildings, restaurants, security facilities, structures)

1.4.1.7 Communications

- Telecommunications network (telephone-exchange stations, cable distribution, power supplies, switching and data-processing centers, buildings, transmission towers, and repeat stations)

- Television/cable network (production stations, transmission facilities, cable distribution, power supply, buildings)
- Wireless/satellite network (satellites, ground-control centers, communication systems, receivers, buildings, services and maintenance facilities)
- Information highway network (computer networks, cable distribution, data-processing hardware/ software, on-line and off-line systems, information sources, buildings, backup and recording mediums)

1.4.2 Examples of infrastructure assets

The road network in the United States exceeds 6 million km (3.9 million mi), as shown in Figure 1.4 by functional classification [Abstract 94]. While the interstate highway system is about 1 percent of the total road network, it carries over 20 percent of the heavy-truck traffic [Choate 81]. As a part of the ISTEA (Intermodal and Surface Transportation Efficiency Act) legislation, Congress has designated about 256,000 km (160,000 mi) of road as the National Highway System (NHS) in 1995. Although the NHS will comprise only 4 percent of the nation's roads, it carries 40 percent of auto travel and 75 percent of truck traffic, and connects 95 percent of the businesses and 90 percent of the households in the United States [FHWA 96]. The national bridge inventory contains 577,000 bridges of which 70 percent were constructed prior to 1935. Many of these bridges are more than 50 years old, and 40 percent are,

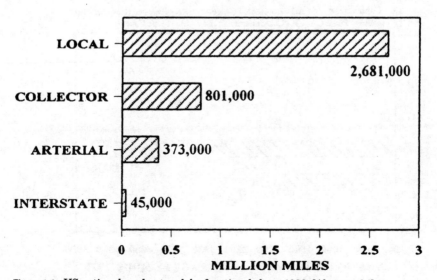

Figure 1.4 US national road network by functional class, 1992. [Abstract 94].

by reasons of their condition or appraisal, eligible candidates for rehabilitation or replacement [Golabi 92].

Figure 1.5 shows the distribution of national transit ways by mode [Abstract 94]. A World Bank study [Faiz 87] provides an overview on the status of main roads in 85 developing countries that have received $9.4 billion in road-assistance programs from the World Bank during the years 1974 to 1985. The aggregate length of the main road systems in these countries is about 1.8 million km, of which slightly more than 1 million km are paved. The main roads carry about 70 to 80 percent of the interurban traffic.

Historically, airports have been reliable and stable and have kept pace with the technological developments in air-traffic-control systems and aircraft fleet. Figure 1.6 shows the advances in aircraft fleet as a result of travel demand [Horonjeff 94]. Average passenger capacity per aircraft was 106 seats in 1969, which grew to 162 seats by 1978 and 181 seats by 1989. New wide-bodied aircraft have a capacity of 300–500 passengers. In 1978, about 250 million passengers in the United States traveled by air, and by 1989 this figure had already reached 445 million [Hamiel 92]. The worldwide airline fleet in 1975–1978 was somewhere in the vicinity of 3,500 to 4,000 aircrafts; by 1989 the fleet grew to 7,441 aircrafts as estimated by Boeing and Airmark.

The railroad network in the United States has been used largely for bulk freight over long distance since the 1960s. Light-rail trains have become effective in some large urban areas. In Europe, Japan, and

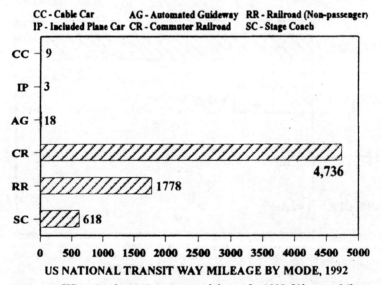

Figure 1.5 US national transitway network by mode, 1992. [Abstract 94].

Figure 1.6 Trends in the gross weight of transport aircraft. [after Horonjeff 94].

other parts of the world, railroads are a popular mode of passenger travel. Major technological advances in speed and equipment have been made in Europe and Japan.

The oil crisis of the early 1970s spurred new exploration of oil and gas all over the globe. Super oil-tanker fleets are being used to transport oil and liquefied gas across oceans. Figure 1.7 shows the US

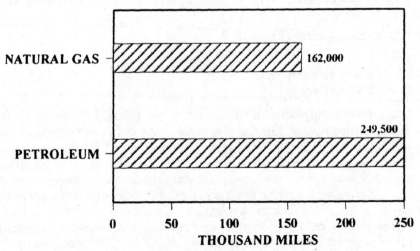

Figure 1.7 US national pipeline network mileage by mode, 1992. [Abstract 94].

pipeline infrastructure mileage of natural gas and petroleum [Abstract 94].

1.4.3 Urban infrastructure

There is a worldwide trend to increased urbanization. About 40 percent of the world population now lives in cities. In the United States, approximately 90 percent of the population inhabits 10 percent of the land [Bragdon 95]. Urban infrastructure and related public services are therefore crucial to our society. For illustration, the following summary of New York City's infrastructure assets is presented [Wagner 84]:

New York City's capital needs are enormous. The extent and variety of New York's infrastructure are extraordinary:

- Forty-seven waterway bridges and 2,057 highway bridges and elevated structures.

- A water-supply system that delivers 1.45 billion gallons of water a day from a reservoir system of 1,956 sq. mi. It delivers water through two tunnels (a third is under construction), 32 billion linear feet of trunk and distribution mains, and 20,000 trunk valves.

- 6,100 mi of sewers, 12 operating water-pollution-control plants and 450 combined sewer-overflow regulators.

- 6,200 mi of paved streets that cover approximately 30 percent of the city's land.

- 6,700 subway cars that ride on 232 mi of track (137 mi underground, 72 mi elevated, and 23 mi open bed) and 4,560 buses.

- An estimated 3,500 acres of landfill, and nine marine transfer stations.

- Over 25,000 acres of parkland.

1.5 Life-cycle Analysis in Planning and Design

The infrastructure construction boom of the 1950s, 1960s, and 1970s in the United States, Europe, and other parts of the world resulted in numerous projects and advances in planning and design practices. Availability and use of computers and telecommunication facilities have helped in these accomplishments. Operational management of the facilities and services has also advanced. However, until recently, much of the infrastructure development did not consider maintenance, future rehabilitation/renovation, and replacement activities in overall planning and costing. Scheduled maintenance has been generally based on experience or on "crisis" urgency. Consequently, the quality of

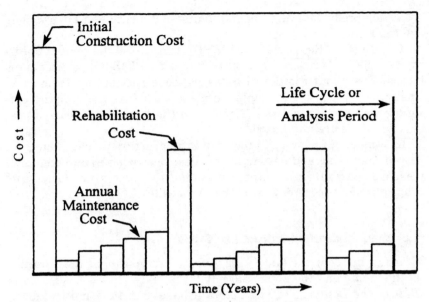

Figure 1.8 Life-cycle cost streams for infrastructure analysis.

service drops because of aging, and in some cases the result can even be sudden failure and loss of life. This points out that life-cycle analysis and ongoing planned monitoring over that life cycle is required.

Figure 1.8 illustrates the life-cycle cost stream of a facility. The condition of the facility can be preserved over its service life if the condition-responsive maintenance and rehabilitation actions are properly timed. This requires the prediction of condition, deterioration, or performance, over the service life as discussed in more detail in Chapter 8. Performance of an infrastructure element or facility is considered good if it performs as designed and provides an acceptable level of service over its intended life. Poor performance indicates that the facility: (1) may have deteriorated faster than predicted; (2) and/or provides an inadequate level of service; and/ or (3) is aged beyond its design life without any major rehabilitation, renovation, or replacement.

One pioneering concept of performance in terms of the present serviceability index (PSI) was developed at the AASHO Road Test [HRB 62] involving highway pavements. Appraisal methods and condition rating procedures for bridges were developed in the early 1970s after the collapse of several structures resulted in loss of life. The concept of maintenance and rehabilitation during the service life of pavement assets was developed in the mid-1970s [Haas 94].

The development of good performance models depends on condition-assessment methods, load and demand data estimates, material behavior prediction, and understanding of climatic and environmental conditions.

The Building Research Board (BRB) reorganized the concept of life-cycle costs in its report *Pay Now or Pay Later* [BRB 91], which stated, "Decisions about a building's design, construction, operation and maintenance can, in principle, be made such that the building performs well over its entire life cycle and the total costs incurred over this life-cycle are minimized."

In summary, the overall process of infrastructure management goes beyond planning and design; it includes construction and acceptance testing, periodic condition assessment, and maintenance and improvement programs during the service life.

1.6   Magnitude of Infrastructure "Crisis"

Starting in about the late 1970s, infrastructure caught the attention of the media and the public, largely initiated by the book *America in Ruins: The Decaying Infrastructure* [Choate 81]. Its condition, decay and aging, and sometimes disastrous effects (including loss of lives and property), made for news headlines. Citizens, media, investors, and public officials became more concerned after hearing the evidence of and the publicity surrounding some critical incidents of sudden collapse and failure of various infrastructure components. The public awareness of these incidents and identification of potential failure areas has led to a perception of reality of an infrastructure crisis in the United States.

1.6.1   Examples of extreme incidents in the infrastructure crisis

1967: Bridge collapses in West Virginia.   The Silver Bridge, on the Ohio River between Point Pleasant, West Virginia and Gallipolis, Ohio-collapsed during rush-hour traffic in 1967. Many vehicles were stopped on the structure for a traffic signal when the instantaneous fracture of an eyebar led to the loss of 46 lives. This disaster was highly publicized and drew attention to the aging condition of the nation's bridges. The US Congress added provisions to the Federal Aid Highway Act of 1968, which required the Secretary of Transportation to establish a national bridge-inspection (NBI) system and to develop a bridge-inspection program [Infra 92]. The standard was issued in April 1971. Since then, bridge inspection has been continuously improved. This was the most influential milestone in the recognition of the infrastructure crisis.

1974: The West Side Highway collapses in New York. A major section of West Side Highway, an important major arterial in New York City, collapsed, resulting in the closing of the highway; no life was lost [Wagner 84].

1982: An 80-year old aqueduct fails in New Jersey. After the failure of an 80-year aqueduct in July 1982, 300,000 residents of Jersey City, New Jersey, had no drinking water for three days [Kwiatkowski 86].

1983: Mianus River Bridge collapses in Connecticut. A section of the Mianus River Bridge, on Interstate 95 in Connecticut, collapsed in 1983, killing three persons and seriously injuring three others [Kwiatkowski 86].

1983: Water main breaks in New York City. Water main breaks in midtown Manhattan closed streets and subways, caused a blackout, and seriously disrupted business [Wagner 84].

1988: A woman falls through a sidewalk grate. In December 1988, New York City tabloids ran wild with the story of a 28-year-old Brooklyn woman who was critically injured when she fell through a weakened sidewalk grate and landed on subway tracks 50 feet below [Perry 89].

1992: Chicago River floods ravage the city. A hole in a 100-year-old freight tunnel system allowed 250 million gallons of Chicago River water to flood the basements of 200 buildings in downtown Chicago, causing $10 billion in damages [Infra 92].

1.6.2 Effects of natural disasters and other incidents on infrastructure

Natural disasters (earthquakes, floods, hurricanes, tornados, volcanic eruptions, ice storms) and serious incidents (fires, riots, terrorist attacks) damage infrastructure heavily and disrupt essential services and business activities, as shown in the following examples:

1992—Hurricane Andrew in South Florida

1993—Bomb Explosion in World Trade Center, New York City

1994—Earthquake in Los Angeles

1995—Earthquake in Kobe, Japan

1995—Bomb explosion in a federal building, Oklahoma City

1.7   Maintenance, Preservation, and Innovation Challenges

A nation's infrastructure represents a sizable asset. The value of this asset in the United States is estimated at $20 trillion in civil infrastructure systems, including all installations that house, transport, transmit, and distribute people, goods, energy, resources, services, and information [NSF 95]. Because of aging, overuse, exposure, misuse, mismanagement, and neglect, many of these systems are deteriorating and becoming more vulnerable to catastrophic failure, particularly when earthquakes, hurricanes, and other natural hazards strike. It would be prohibitively costly and disruptive to replace these vast networks. They must instead be renewed in an intelligent manner, which includes the prudent and effective use of our economic, material, and human resources, and which focuses on optimizing the performance of both individual subsystems and of the civil infrastructure systems complex as a whole.

In the public-works arena, these systems have evolved in a piecemeal fashion, with new extensions grafted onto existing systems and designs often governed by expediency and low construction costs rather than true life-cycle costs. We have inherited a complex network of systems comprised of subsystems with wide variations in age and functionality. How systems interact is often poorly assessed, and maintenance has often been inadequate [NSF 95].

1.7.1   Preservation and repair estimates

Expenditures on infrastructure preservation and major repair are about 20 to 40 percent of total new construction in the United States, more than $80 billion annually. The ways in which we allocate and manage these resources influence the returns we realize. Inefficiencies of only a few percentage points will represent substantial losses.

The Congressional Budget Office has estimated that for the 756 urban water-supply systems in the United States, between $63 and $100 billion will be needed by the year 2000 to replace all water mains over 90 years of age and to replace other mains as necessary. Extrapolations to all community water systems (adjusted for population variations) suggest that total replacement and rehabilitation needs for all communities could run as high as $160 billion by the year 2000 [O'Day 84].

As reported in *Fortune*, the National Council on Public Works Improvement, which was created in 1984 to assess the state of America's public works (the report card), concluded in its final report to Congress in 1988 [Perry 89]:

While America's infrastructure is not in ruins, it is inadequate to sustain future economic growth. America has to face its needs even as the federal government is moving out of the infrastructure business. Its last heroic public works project, the interstate highway system, will be completed in 1992. Preoccupied with an unyielding budget deficit, Uncle Sam is doling out less and less money for infrastructure. Heywood Sanders estimates that of the $40.5 billion spent to expand highways, airports, sewers, mass transit, and waterworks last year, 46% came from federal funds, down from 54% in 1985. State and municipal governments are stuck with the rest.

1.7.2 Infrastructure maintenance problems—national perspective

A report on the problems facing the preservation of the infrastructure [Perry 89] asks, "What needs fixing? Many of the sewers, bridges, and water systems in America's older cities, built around the turn of the century, are now in disrepair. The Federal Highway Administration's latest report classifies 23 percent of the bridges included in the national bridge inventory in the U. S. as structurally deficient. They are either closed or restricted to light traffic. Poor roads cost American motorists billions of dollars in wasted fuel, added tire wear, and extra vehicle repair. Breakage does not always imply decay. The financing of highway pavement preservation has been improving since 1982, when Congress raised the federal gas tax by five cents per gallon."

The following excerpt from *Fortune* (1989) gives more examples of maintenance programs needed to improve the infrastructure and related services [Perry 89]:

> Still, travelers feel that the roads and airports they use are overwhelmed by congestion. According to the Federal Highway Administration, 65 percent of the traffic at peak travel times on interstates in urban areas moves at an average speed of less than 35 miles an hour, up from 54 percent in 1983. Though the U.S. has 16,300 public airports—more than the rest of the world combined—no new commercial airports have been built since 1974 (Dallas-Fort Worth) to 1990. Since then the number of passengers carried has more than doubled to 479 million. By 1993 up to one-half of the country's landfills, which collect 95% of the 450,000 tons of solid waste Americans generate each day, will fill up and be closed. Partly in an effort to promote alternatives, such as recycling and incineration, cities in some parts of the U.S. are doubling or tripling the price of waste disposal.

1.7.3 Airport capacity problems

In Europe, 24 airports risk becoming frequency-limited by the turn of the century. These airports today handle 55 percent of all commercial air-transport movements in Europe. Their self-declared present maximum runway capacity is in the order of 4.6 million movements per year. In 1988, these airports handled almost 4 million movements. This leaves only marginal opportunities for future growth in flight frequency. Assuming a 20 percent capacity improvement by early in the next century, achieved by a better organization of resources, frequency could increase 1.9 percent annually [Jost 92].

1.7.4   Road infrastructure maintenance in the United Kingdom

According to a World Bank study [Faiz 87], the National Development Office reported that the gross replacement value of the entire UK road stock in 1982, including the plant and equipment owned by central and local road authorities, was estimated at about £38 billion. The deficiencies in the system as compared with a brand-new system were estimated at about £10 billion or 26 percent of the replacement value.

1.7.5   Maintenance of urban infrastructure

The extensive and extraordinary diversity and size of the urban infrastructure of New York City has been described earlier. In parallel there are clear indications of neglect to maintain and preserve the condition of these facilities, as follows [Wagner 84]:

> In 1978, the Koch administration found a pattern of neglect almost frightening in its extent. The desirable rate for repaving streets is once every 25 to 50 years, depending on usage. In 1978 the city was repaving streets at the rate of once every 200 years. Engineers say a water main will require replacement probably every 100 years. In 1978 we were replacing water mains at the rate of once every 296 years, in 1978 the state found 135 waterway bridges and highway structures to be in poor condition and requiring major reconstruction. The same pattern would apply to all other parts of the city's physical plant.

1.7.6   Infrastructure research and development efforts

In 1987, the Congressional Office of Technology Assessment reported that Japan outspent the US on research and development for public works by a factor of 30. Europe outspent the US in this area by a factor of eight. Furthermore, our civil infrastructure has not benefited much from either US or foreign advances in technology. A combination of the highly competitive nature of the infrastructure construction industry, the difficulty of disseminating new technologies, negative incentives for adopting new technologies, and excessive litigation has impeded technological advances in the industry [NSF 95].

1.8   Infrastructure Management—
An Integrated Approach

1.8.1   Reasons for infrastructure deterioration

Infrastructure problems have compounded over recent decades for several reasons: (1) the underinvestment in public works programs; (2) the lack of good management systems for infrastructure; (3) failure to recognize the importance to the future economy of maintaining a sound physical infrastructure; (4) cutbacks that have slashed public works budgets; (5) failure to replace the infrastructure as fast as it wears out; (6) failure to realize that lack of physical infrastructure seriously impacts the level and types of services government can provide to their citizens; (7) tendency by national, state, and local officials to defer the maintenance of public infrastructure; and (8) increased costs to taxpayers to repair and rebuild the obsolescent public infrastructure.

Aschauer has shown that the productivity (i.e., output per unit of private capital and labor) is positively related to government spending on infrastructure, including roads [Aschauer 89]. Analyzing data from the United States for the period 1949 to 1985, he observed that underinvestment in infrastructure started in about 1968, and the effects of deterioration became evident half a decade later, when a productivity slump began in the United States [Queiroz 92].

1.8.2   Overall framework for infrastructure management

The past underinvestment in infrastructure maintenance and the lack of overall system management principles point out the need for better management and financing approaches. In turn, it is essential that available funds be spent in a cost-effective and timely way. The thrust should be on the preservation of the condition and value of assets. An example of leadership in this regard is the Intermodal Surface Transportation Efficiency Act (ISTEA) of 1991, which specified six management systems for State DOT's (pavement, bridge, safety, congestion, transit, and intermodal). As well, the Federal Highway Administration (FHWA) has been active in supporting efforts to improve bridge-management system concepts [Hudson 87, Golabi 92].

An overall framework for any infrastructure management system is illustrated in Figure 1.9. The major point of the diagram is that management can be divided into two distinct but closely integrated levels: program/network/systemwide and project/section. Key elements of the overall framework for infrastructure management are

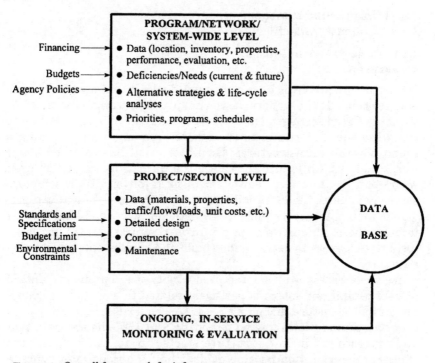

Figure 1.9 Overall framework for infrastructure management.

ongoing, in-service monitoring and evaluation, and a database. Each of the two levels of management must consider exogenous elements over which little or no control may exist, such as financing, budgets, and agency policies for the network level, and standards and specifications, budget limits, and environmental constraints for the project level.

Comprehensive infrastructure assets can be found in large urban areas, as illustrated in Figure 1.10. Municipal infrastructure includes many facilities providing transportation services, utilities, health and education services, public buildings, parks, and other recreational areas. The exact scope of an infrastructure management system for a specific community and jurisdiction will depend on the network size and extent of services to be covered.

Figure 1.10   An illustration of infrastructure assets in a large metropolitan area.

References

[Abstract 91]   Bureau of Census, *Statistical Abstracts of the United States 1991: The National Data Book,* U.S. Department of Commerce, Washington, D.C., 1991.

[Abstract 94]   Bureau of Census, *Statistical Abstracts of the United States 1994: The National Data Book,* U.S. Department of Commerce, Washington, D.C., 1994.

[AGCA 82]   Associated General Contractors of America, *Our Fractured Framework: Why America Must Rebuild,* Washington, D.C., 1982.

[Aschauer 89]   D. A. Aschauer, "Infrastructure Expenditures and Macro Trends," *Proceedings of the Africa Infrastructure Symposium,* The World Bank, Washington, D.C., 1989.

[Bragdon 95]   C. R. Bragdon, "Intermodal Transportation Planning for the 21st Century, a New Paradigm: An Urgent Call for Action," presented at the 74th Annual Meeting of the Transportation Research Board, National Research Council, Washington, D.C., 1995.

[BRB 91]   Building Research Board, *Pay Now or Pay Later,* National Research Council, Washington, D.C., 1991.

[Choate 81]   P. Choate, and S. Walter, *America in Ruins: The Decaying Infrastructure,* Duke Press, Durham, N.C., 1981.

[Colorado 84]   University of Colorado at Boulder, *Hard Choices: A Report on the Increasing Gap between America's Infrastructure Needs and Our Ability to Pay for Them,* Denver, 1984.

[DOT 94]   Department of Transportation, *National Transportation Statistics,* U.S. Department of Transportation, Washington, D.C., 1994.

[Faiz 87]   A. Faiz, C. Harral, and F. Johansen, "State of the Road Networks in Developing Countries and a Country Typology of Response Measures," in

Transportation Research Record 1128, Transportation Research Board, National Research Council, Washington, D.C., 1987, pp. 1–17.

[FHWA 91]   Federal Highway Administration, *Highway Statistics,* U.S. Department of Transportation, Washington, D.C., 1991.

[FHWA 96]   Federal Highway Administration, "Public Roads," A Journal of Highway Research and Development, U.S. Department of Transportation, Washington, D.C., Winter 1996.

[Golabi 92]   K. Golabi, P. Thompson, and W. A. Hyman, *Pontis Technical Manual,* prepared for the Federal Highway Administration, January 1992.

[Grigg 88]   N. S. Grigg, *Infrastructure Engineering and Management,* John Wiley and Sons, New York, 1988.

[Haas 94]   R. Haas, W. R. Hudson, and J. P. Zaniewski, *Modern Pavement Management,* Krieger Publishing Company, Malabar, Fla., 1994.

[Hamiel 92]   J. Hamiel, "Changing Trends at U.S. Airports as a Part of the International Scene," *Transportation Research Circular 393—Trends and Issues in International Aviation,* Transportation Research Board, Washington, D.C., pp. 21–24.

[Horonjeff 94]   R. Horonjeff, and F. X. McKelvey, *Planning and Design of Airports,* McGraw-Hill, New York, 1994.

[HRB 62]   Highway Research Board, "The AASHO Road Test: Report 5-Pavement Research," *HRB Special Report 61-E,* National Research Council, Washington D.C., 1962.

[Hudson 87]   S. W. Hudson, R. F. Carmichael III, L. O. Moser, W. R. Hudson, and W. J. Wilkes, "Bridge Management Systems," *NCHRP Report 300,* National Cooperative Highway Research Program, Transportation Research Board, National Research Council, Washington, D.C., 1987.

[Infra 92]   "A Century of Building America," *Infrastructure,* July 1992.

[Jost 92]   P. Jost, "Forecasting Traffic and Capacity Growth in Europe," *Transportation Research Circular 393—Trends and Issues in International Aviation,* Transportation Research Board, Washington, D.C., pp. 5–15.

[Kwiatkowski 86]   V. F. Kwiatkowski, "Infrastructure Assets: An Assessment of User Needs and Recommendations for Financial Reporting," Ph.D. thesis, University of Kentucky, 1986.

[NSF 94]   National Science Foundation, *Civil Infrastructure Systems Research,* Washington, D.C., 1994.

[NSF 95]   National Science Foundation, *Civil Infrastructure Systems—An Integrated Research Program,* Washington, D.C., 1995.

[O'Day 84]   D. K. O'Day, "Aging Water Supply Systems: Repair or Replace," *Infrastructure—Maintenance and Repair of Public Works, Annals of the New York Academy of Sciences,* Vol. 431, Dec. 1984, pp. 241–258.

[Perry 89]   N. J. Perry, "Good News about Infrastructure," *Fortune,* April 10, 1989, pp. 94–99.

[Queiroz 92]   C. Queiroz, and S. Gautam, "Road Infrastructure and Economic Development—Some Economic Indicators," *Working Papers WPS 921,* Western African Department and Infrastructure and Urban Development, The World Bank, Washington, D.C., June 1992.

[Queiroz 94]   C. Queiroz, R. Haas, and Y. Cai, "National Economic Development and Prosperity Related to Paved Road Infrastructure," *Transportation Research Record 1455,* Transportation Research Board, National Research Council, Washington, D.C., 1994, pp. 147–152.

[Stone 74]   D. C. Stone, *Professional Education in Public Works / Environmental Engineering and Administration,* American Public Works Association, Chicago, 1974.

[Veldhuis 92]   J. Veldhuis, "Impact of Liberalization on European Airports," *Transportation Research Circular 393—Trends and Issues in International Aviation,* Transportation Research Board, Washington, D.C., pp. 25–30.

[Wagner 84]   R. F. Wagner, Jr., "Infrastructure Issues Facing the City of New York," *Infrastructure—Maintenance and Repair of Public Works, Annals of the New York Academy of Sciences,* Vol. 431, Dec. 1984, pp. 21–26.

[World 90]   The World Bank, *World Development Report 1990,* Washington, D.C., June 1990.

2

Framework for Infrastructure Management

2.1 Background

Infrastructure represents the physical assets and their related actions that serve the economic and social needs of the public. As Sullivan stated at the 1983 New York Conference on Infrastructure: Maintenance and Repair of Public Works [Sullivan 84], "There is nothing new about infrastructure. It is simply the fundamental basis on which this country was built—public works." It is also reasonable to say that infrastructure management is not new. Management decisions must be made every day in public works agencies for project development, financing, construction, and maintenance. The exact style and type of management organization and activity often depends more on historical practice of the individual agency than on need or reality. Unfortunately, public agencies have not realized the importance of performance evaluation, maintenance programming, or other important keys to successful asset management.

In the past, managing infrastructure has not always been systematic. There is a need to change the way we do business. The infrastructure management system (IMS) concepts introduced in this book can provide a framework and methods to link all phases involved with the provision of infrastructure.

The terms *infrastructure, management,* and *system* mean different things to different people. Therefore, it is important to identify key terminology and provide technical definitions so that a uniform, common vocabulary exists for use in the infrastructure management field. We need to build on success in other fields and apply new technology where appropriate and needed.

2.1.1 Terminology and definitions

2.1.1.1 System. The word *system* has been appropriated for many purposes, such as circulatory system, drainage system, sprinkler system, highway system, parking system, etc. The dictionary says that a system is a regularly interacting or interdependent group of items comprising a unified whole. This last definition will be used herein.

2.1.1.2 Infrastructure. The word *infrastructure,* as used in this book, refers to physical systems or facilities that provide essential public services, such as transportation, utilities (water, gas, electricity), energy, telecommunications, waste disposal, park lands, sports, recreational buildings, and housing facilities. Infrastructure can also include the management and human resources associated with providing a physical facility. Infrastructure consisting of physical systems can be owned and managed by either or both public agencies and private enterprises.

2.1.1.3 Management. The word *management* has diverse meanings. To some it means to administer, to others it means control, and to still others it means coordinating the various elements of some unit or system. The dictionary definition of management is "the act or art of managing," or, less circularly, "the judicious use of means to accomplish an end." In this book, management means the coordination and judicious use of means and tools, such as funding and economic analysis to optimize output or accomplish a goal of infrastructure operation.

2.1.1.4 Infrastructure management. *Infrastructure management* includes the systematic, coordinated planning and programming of investments or expenditures, design, construction, maintenance, operation, and in-service evaluation of physical facilities. It is a broad process, covering those activities involved in providing and maintaining infrastructure at a level of service acceptable to the public or owners. These activities range from initial information acquisition to the planning, programming, and execution of new construction, maintenance, rehabilitation, and renovation; from the details of individual project design and construction to periodic in-service monitoring and evaluation.

2.1.1.5 Infrastructure management system (IMS). An *infrastructure management system* (IMS) consists of the operational package (methods, procedures, data, software, policies, decisions, etc.) that links and enables the carrying out of all the activities involved in infrastructure management.

2.1.2 Ideal infrastructure management system

An ideal infrastructure management system would coordinate and enable the execution of *all* activities so that optimum use is made of the funds available while maximizing the performance and preservation of assets and provision of services. It would serve all management levels in the organization (public or private), and would be structured to be adaptable to all of infrastructure. In other words, it would be general in scope and incorporate particular models, methods, and procedures needed for specific types of infrastructure. For example, the concept of pavement management, initiated in the late 1960s, was later generalized to bridges and underground utilities.

2.2 Key Issues for Infrastructure Management

The infrastructure crisis was identified and discussed in Chapter 1. Public attention was certainly caught in the 1980s by such headlines as "America in Ruins"and "Crumbling Infrastructure." Huge needs and lack of funds to maintain and improve the infrastructure are often cited as the cause of this problem. While cost is a factor, a major problem has been the lack of a comprehensive approach to managing the infrastructure. This involves the following major issues.

2.2.1 Infrastructure decaying/aging.
The following key points are related to infrastructure decay and aging:

- The condition and level of service of infrastructure has deteriorated through aging and usage.
- Some infrastructure components have failed due to natural disasters, such as earthquakes and floods.
- Historically, design processes have not given adequate consideration to environmental effects and their interaction with loads and material variability.

2.2.2 Lack of rational maintenance, preservation, and renovation programs.
Lack of rational maintenance, preservation, and renovation programs is related to the following observations:

- Generally, past design practices were geared toward producing physical systems that would last a given design life with no maintenance or future preservation treatments, such as renovation, considered.

- Routine maintenance was considered to be the responsibility of a maintenance group. The condition and level of service concepts suggested that facilities fail and are replaced; alternative actions were not considered.

- "Ad-hoc" maintenance practices in response to public complaints, emergency situations, and catastrophic failures are not adequate to sustain healthy infrastructure.

- Changes in use and inability to accurately predict future loads and service requirements have caused problems.

- Inadequate attention has been given to performance-prediction models.

These factors define a need for appropriate management of infrastructure.

2.2.3 Scarcity of financing resources. Traditionally, the federal government has financed most of the national public works infrastructure, while states and local agencies have financed infrastructure related to their jurisdictions. However, the accumulated federal budget deficit has been steadily rising from $195 billion in 1983 to $600 billion in 1994. There is strong pressure to cut federal spending and bring the deficit under control within 7 to 10 years. At the same time, competing demands make the federal budget a combination of solemn and deeply felt commitments to people, high-priority emergencies, and absolutely essential expenditures. The common justifications for public spending have extremely strong political appeal [Mathiasen 84].

Innovations are needed to identify financing resources. States and local governments need more flexibility in using the available funds. Cost-effective solutions and better management of funds are essential. Incidents like the bankruptcy of Orange County, California, in 1994 can shatter investor confidence in public works agencies. Better analytical tools are needed to improve priority programming. These considerations make it essential to adequately educate and train engineers and decision makers for cost-effective infrastructure management. This book is intended to provide the basic principles and techniques for improved infrastructure management.

2.2.4 Inadequate financial reporting. Infrastructure inventory and monitoring of costs are important issues that have not been fully recognized in government accounting and financial reporting procedures. As a consequence, needed infrastructure information is not always available to decision makers. The US Governmental Accounting Standards Board (GASB), established in 1984, initiated a first comprehensive look at accounting and reporting of infrastructure/fixed assets. Kwiatkowski has conducted research to assess the needs of selected users of financial information about infrastructure and to recommend procedures for financial reporting of infrastructure asset information [Kwiatkowski 86].

2.3 Application of Systems Methodology

A systems methodology can address various modeling and analytical issues related to infrastructure management. It has been successfully used in areas such as pavement and bridge management [Haas 78, Haas 94, Hudson 87].

Systems engineering comprises a body of knowledge that has been developed for the efficient planning, design, and implementation of new systems, and for structuring the state of knowledge on an existing system or modeling its operation. There are two main, interrelated applications of systems methodology:

1. The framing or structuring of a problem, body of knowledge, or process.
2. The use of analytical tools for actually modeling and solving the problem or for incorporation in the process.

These uses are complementary and interrelated; one is insufficient without the other. The framing of a problem is usually too generalized by itself to achieve a useful operational solution, whereas the application of analytical techniques to an inadequately structured problem may result in an inappropriate solution [Stark 72].

The structure or framework of any problem-solving process should provide for systematic consideration of all the technical, economic, social, and political factors of interest. Moreover, it should be a logical simulation of the progression of activities involved in efficiently solving or dealing with a problem. A framework for infrastructure management is subsequently presented in this chapter. The following discussion concentrates on a summary of some analytical tools and some precautions. These are applicable within the general systems approach shown in Figure 2.1.

2.3.1 Some analytical tools

The structuring alone of a problem is usually too general to be useful in finding an operational solution. Moreover, the application of analytical techniques, no matter how elegant or complex, to an inadequately structured problem will likely result in an inappropriate solu-

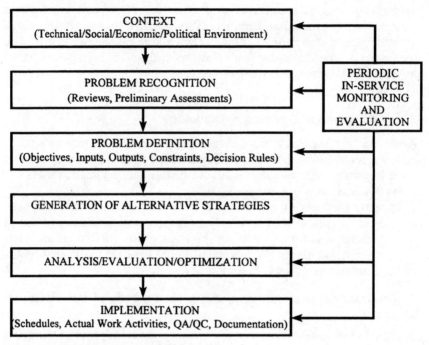

Figure 2.1   General systems methodology approach.

tion. In other words, analytical techniques for solving problems have maximum usefulness when the problems are well formulated or structured.

This section provides a "catalogue" of some of the more widely used techniques, tools, and models. Use of such techniques should facilitate reaching a decision on objective basis. Their use depends largely on the available knowledge of the outputs of the system, which can be classified as follows:

1. Certainty, where definite outputs are assumed for each alternative strategy (i.e., deterministic type of problem).

2. Risk, where any one of several outputs, each with an identified or estimated probability, can occur for each alternative strategy.

3. Uncertainty, where the outputs are not known for the alternative strategies; thus probabilities cannot be assigned.

A majority of engineering practice has treated problems in terms of decisions with certainty, because of convenience and the need to deal with many variables. However, there is considerable need to incorporate risk concepts into practice. Where practical problems are too complex for symbolic representation, Stark suggests that they may be modeled on an analogue or a scalar basis. Alternatively, it is possible to "force" a solution by experimentation, gaming, or simulation for some types of problems [Stark 72].

The optimization problems for finding cost-effective solutions can be solved by several approaches:

- Mathematical programming, which produces an exact solution
- Heuristics, which are more common for large problems and produce suboptimal solutions
- Probabilistic approaches, which are based on random selection, biased sampling, and Monte Carlo simulation
- Graphical solutions

One of the most widely applied and useful classes of systems models involves linear programming. These techniques have been used in everything from construction to petroleum refinery operations because they are well suited to allocation-type problems [Gass 64]. A typical problem for linear programming application might involve the determination of how much of each type of material a materials supplier should produce, given production capacity, the number and capacities of trucks, available materials and their costs, etc. There are several variations of linear programming models and several methods of solution, including parametric linear programming, integer linear programming, and piecewise linear programming. The latter is used to reduce a nonlinear problem to approximate linear form in the area of interest.

Nonlinear methods range from the so-called classical use of differential calculus, Lagrange multipliers (and their extension to nonnegativity conditions and inequality constraints), and geometric programming to the iterative search techniques [Künzi 66]. These latter techniques are often applicable where more rigorous methods are impractical. There are some types of nonlinear problems not easily solved by analytical techniques that may lend themselves quite well to graphical solutions. For example, a simple graphical solution has been shown to be applicable to a construction problem involving a discontinuous cost function [Haas 73].

Problems involving multistage decisions can be represented as a sequence of single-stage problems. These can be successively solved by dynamic programming methods [Dreyfus 65]. Each single variable or single-stage problem that is involved can be handled by the particular optimization technique that is applicable to that problem. These techniques are not dependent on each other from stage to stage and can range from differential calculus to linear programming.

Combinatorial-type problems are often well suited to dynamic programming. A typical example is an aggregate producer with several mobile crushers and several sources of raw materials that wants to determine how many crushers should be assigned to each site for a given profit margin.

Random and queuing models have a wide range of applicability to systems problems, and there is a lot of literature available. One class of models involves Monte Carlo methods, which are useful when adequate analytical models are not available. These probabilistic methods require distribution functions for the variables. There are also a large range of problems to which reliability, random walk, and Markov chain techniques can be applied. The latter can be used to extend stochastic and chance-constrained programming models. Queuing models have been used extensively in engineering, including various air terminal operations, traffic facility operations, rail operations, and canal operations.

Many problems involve the allocation and scheduling of personnel, equipment, money, and materials. Several project management techniques have found widespread use for these types of problems, including sequencing, routing, and scheduling.

This section has noted only a few of the many analytical tools that have potential applicability to various aspects of infrastructure management. Those desiring more in-depth information may consult some of the many references cited and others available in libraries.

2.3.2 Some precautions in application

The general systems approach of Figure 2.1 models the logical, systematic pattern that is used by efficient problem solvers. It must, however, be used with recognition that there are limitations.

1. Successful application inherently depends on the capabilities of the people involved. The approach is not a substitute for poor judgment or poor engineering.

2. The point of view of the individual or agency involved must be clearly recognized and identified. Otherwise, confusion and contradiction can result. For example, a materials-processing problem for a public-works project might be viewed differently by the contractor than by the government agency involved. They may have competing objectives, and they will have different constraints.

3. The components or extent of the system under consideration should be clearly identified. For example, the term "parking system" might mean the actual physical parts of the parking lot, such as pavement, curbs, and gates, to one person; to another it might

mean the method used to operate the facility; and to still another person it might mean a combination of the two.

4. A fourth point concerns the inherent danger of generating precise solutions to an imperfectly understood problem. That is, the problem has been recognized but not yet rigorously defined. It is, of course, common to perceive some general solutions in the problem-recognition phase, but these may be inadequate or incomplete if the problem solver does not go further to define the problem.

2.4   Development of Infrastructure Management System (IMS)

2.4.1   IMS process

From the systems-engineering perspective, a system consists of a set of interacting components that are affected by certain exogenous factors or inputs. In a physical airport-facility system the set of mutually interacting components usually includes:

1. A ground access facility connecting the airport to the nearest city

2. Parking structures

3. An air-traffic control tower

4. A terminal building

5. Jetways

6. Apron(s)

7. Taxiway(s)

8. Runway(s)

Each of these components involves different types of construction. For example, all paved surfaces serving vehicular traffic and aircraft operations usually consist of pavement (with surface layer, base and subbase layers, and subgrade), shoulder or sidewalk, and other appurtenances. The exogenous or external factors that affect the building structures and paved surfaces are age, traffic, environment, material degradation, disasters/accidents, and maintenance actions. Maintenance is carried out to preserve the functionality and structural integrity by reducing the rate of deterioration from the impacts of the traffic and environment inputs.

An *infrastructure management system,* on the other hand, consists of such mutually interacting components as planning, programming, design, construction, maintenance and renovation, and evaluation. The overall framework for such a system is illustrated in Figure 1.9. Exogenous factors affecting an *infrastructure management system* include budgets, decision criteria and maintenance policies, and non-quantifiable agency policies, including political climate.

An ideal infrastructure management system (IMS) would help provide and maintain comfortable, safe and economical physical infrastructure systems and related services at acceptable standards to the public, within the available funds. It would recognize the consequences of unwanted delays in the implementation of maintenance programs and assist decision makers in spending the available funds cost-effectively. The minimum requirements of such a system would include adaptability, efficient operation, practicality, quantitatively based decision-making support, and good information feedback. There is no ideal single IMS that is best for all agencies. Every agency presents a unique situation with specific needs for various different components of the physical infrastructure system. Any existing decision support or management system for some infrastructure components should be carefully integrated if an overall infrastructure management system is to be developed. Each agency must establish objectives and goals for an infrastructure management system. These objectives and goals may vary considerably, depending on the jurisdiction of the agency at the national, state, and regional levels, and local levels of government or private enterprises.

The scope of an infrastructure management system is dependent on the extent and size of the physical components of infrastructure systems that an agency is responsible for. In the case of a *municipal infrastructure management system,* all public works infrastructure may be included in the scope. That generally implies city-street infrastructure, water-supply and sewer infrastructure, electricity and gas-supply infrastructure, mass-transit infrastructure, airport infrastructure, and coliseum/convention hall/school and recreational facility infrastructure. There is a network or portfolio of each category of physical infrastructure system.

2.4.2 The network and project levels of infrastructure management

Infrastructure management has two basic working or operational levels: network and project, as illustrated previously in Figure 1.9. Figure 2.2 expands the major activities occurring at each level for highway/street infrastructure. These activities are discussed in more detail in subsequent chapters.

Network-level management has as its primary purpose the development of a priority program and schedule of work, within overall budget constraints. Project-level work thus comes on-stream at the appropriate time in the schedule, and represents the actual physical implementation of network decisions.

TRANSPORTATION, HIGHWAY / STREET SYSTEM MANAGEMENT

NETWORK MANAGEMENT LEVEL

- Sectioning, Data Acquisition (field data on roughness, surface distress, structural adequacy, surface friction, geometrics, etc., plus traffic, costs and other data) and Data Processing
- Criteria for Minimum Acceptable Serviceability, Maximum Surface Distress, Minimum Structural Adequacy, etc.
- Application of Deterioration Prediction Models
- Determination of Now Needs and Future Needs, Evaluation of Options and Budget Requirements
- Identification of Alternatives, Development of Priority Programs and Schedule of Work (rehabilitation, maintenance, new construction)

PROJECT MANAGEMENT LEVEL

- Subsectioning, Detailed Field/Lab and Other Data on Scheduled Projects, Data Processing
- Technical (Predicting Deterioration) and Economic Analysis of Within-Project Alternatives
- Selection of Best Alternative; Detailed Quantities, Costs, Schedules
- Implementation (construction, periodic maintenance)

Periodic Updating of Data; Development and Application of New Methods and Procedures

Figure 2.2 Basic operating levels of pavement management and major activities. [Haas 94].

2.4.3 Influence levels of IMS components

Four of the major components or subsystems (planning, design, construction, and maintenance) have important but changing impacts in terms of a "level-of-influence" concept. This concept (Figure 2.3), which has been used in sectors of industry, such as manufacturing and heavy industrial construction [Barrie 78], shows how the potential effect on the total life-cycle cost of a project decreases as the project evolves.

The lower portion of Figure 2.3 presents a simplified picture, in bar-chart form, of the length of time each major component acts over the life of the infrastructure facility. The upper portion shows plots of increasing expenditures and decreasing influence over the infrastructure life. Expenditures during the planning phase are relatively small compared with the total cost. Similarly, the capital costs for construction are a fraction of the operating and maintenance costs associated

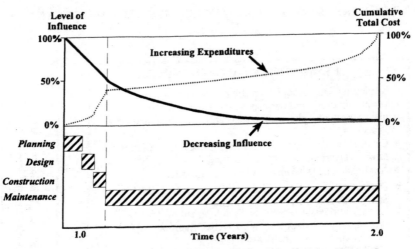

Figure 2.3   Influence levels of IMS subsystems on the total costs. [after Haas 94].

with service life. However, the decisions and commitments made during the early phases of a project have far greater relative influence on later-required expenditures than some of the later activities.

At the beginning of a project, the agency controls all (100 percent influence) factors in determining future expenditures. The question is, to build or not to build? A decision not to build requires no future expenditure for the project. A decision to build requires more decision making, but initially at a very broad level. For example, in the case of a highway, should it be a flexible pavement or a rigid pavement, and, if rigid, with joints or continuously reinforced? How thick should it be and with what kind of materials? Once decisions are firm and commitments are made, the further level of influence of future actions on the future project costs will decrease.

In the same manner, decisions made during construction, even within the remaining level of influence, can greatly impact the costs of maintaining or rehabilitating the infrastructure. For example, lack of quality control or substitution of inferior materials may save a few dollars in construction costs, but the extra maintenance costs and user costs due to more frequent maintenance activities may consume those "savings" several times over.

With construction completed, attention is now given to maintaining the existing infrastructure at a satisfactory level. The level-of-influence concept can also be applied to the subsystems of a maintenance management system (MMS). Expenditures during the planning phases of rehabilitation and renovation are relatively small compared to the total maintenance cost during the service life of the facility.

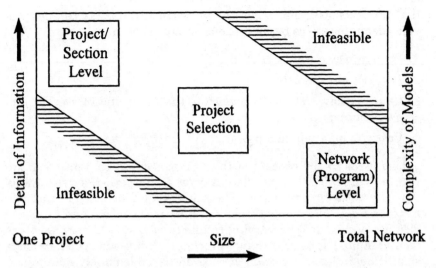

Figure 2.4 Information detail and complexity of models for a three-level IMS.

However, the decisions and commitments made during early phases of a rehabilitation and renovation project have far greater relative influence on what other maintenance expenditures and user costs will be required later.

Some define a three-level concept of IMS that represents the situation that exists in a number of agencies. Also, the terminology sometimes overlaps; for example, in some papers when "project level" is mentioned, the "project-selection level" actually is meant, and in other cases when the "network-level" is mentioned, it is the "program" level that is involved.

This three-level concept is illustrated in Figure 2.4. The lower-left triangle represents an area of unreliability because too little information is available for models at the project level, and the upper-right triangle is an area infeasible for modeling due to the size and complexity of the required models.

2.5 Life-cycle Analysis Concept

The physical infrastructure facilities discussed in Chapter 1 are generally fixed assets. From the design and analysis point of view, some finite number of years of the design life/analysis period is associated with each component of infrastructure. In reality, the public and users expect the infrastructure to provide a particular service forever, unless a catastrophic failure occurs or the area is uninhabited. However, the responsible agency managers and decision makers know that there comes a time when the infrastructure facility can not provide adequate service because of one or more reasons:

1. Structurally unsafe
2. Functionally obsolete
3. Causes delay and inconvenience to the users due to overuse and overdemand
4. Costly to maintain and preserve

This leads to the concept of the "service life" of an infrastructure within a life-cycle. Unlike the design/analysis period, service life is typically not a single number. The same type of facility (for example, a steel bridge) may have a wide variation in its initial and total service life because of the varying influence of traffic history, environmental inputs, and maintenance practices. Maintenance history has a significant influence on total service life. An adequately maintained facility will have a better probability of extended service life, as compared to a poorly maintained facility. A good infrastructure management system recognizes the importance of service-life analysis, including agency costs (for construction, maintenance, rehabilitation, and renovation/replacement) as well as user costs and benefits (Figure 1.8).

The concept of service life is based upon the physical service life, as contrasted to a social/economic service-life estimate, which may be different. Throughout this book, physical service life is used for infrastructure management. Some typical expectations of infrastructure service life are listed in Table 2.1. A detailed discussion on methods to estimate service life is provided in the next chapter.

TABLE 2.1   Typical Expectations of Infrastructure Service Life

Infrastructure Facility and Components		Expected Service Life
Airports	Buildings/Structures	Up to 150 years
	Runways/Taxiways/Aprons	Up to 50 years
Bridges	Decks	Up to 50 years
	Substructure/Superstructure	Up to 125 years
Tunnels	(For auto traffic, water)	Up to 200 years
Public Buildings and Sports Complexes	(Concrete/Steel Construction)	Up to 300 years
Electricity Transmission/ Telephone Lines		Up to 400 years

References

[Barrie 78] D. Barrie, and B. Paulson, *Professional Construction Management,* McGraw-Hill, New York, 1978.

[Dreyfus 65] F. E. Dreyfus, *Dynamic Programming and the Calculus of Variations,* Academic Press, San Diego, Calif., 1965.

[Gass 64] S. I. Gass, *Linear Programming Methods and Applications,* McGraw-Hill, New York, 1964.

[Haas 73] R. C. G. Haas, W. A. McLaughlin, and V. K. Handa, "Systems Methodology Applied to Construction," *Proceedings, National Conference on Urban Engineering Terrain Problems,* Montreal, May 1973.

[Haas 78] R. C. G. Haas, and W. R. Hudson, *Pavement Management Systems,* McGraw-Hill, New York, 1978.

[Haas 94] R. Haas, W. R. Hudson, and J. P. Zaniewski, *Modern Pavement Management,* Krieger Publishing Company, Malabar, Fla., 1994.

[Hudson 87] S. W. Hudson, R. F. Carmichael III, L. O. Moser, W. R. Hudson, and W. J. Wilkes, "Bridge Management Systems," *NCHRP Report 300,* National Cooperative Highway Research Program, Transportation Research Board, National Research Council, Washington, D.C., 1987.

[Kwiatkowski 86] V. F. Kwiatkowski, "Infrastructure Assets: An Assessment of User Needs and Recommendations for Financial Reporting," Ph.D. thesis, University of Kentucky, 1986.

[K,ünzi 66] H. B. Künzi, W. Krelle, and W. Oettli, *Nonlinear Programming,* Blaisdell, 1966.

[Mathiasen 84] D. G. Mathiasen, "Federal Perspectives on Public Works Infrastructure: One Person's View from the Executive Branch," *Infrastructure— Maintenance and Repair of Public Works, Annals of the New York Academy of Sciences,* Vol. 431, Dec. 1984, pp. 5–11.

[Stark 72] R. M. Stark, and R. L. Nicholls, *Mathematical Foundations for Design: Civil Engineering Systems,* McGraw-Hill, New York, 1972.

[Sullivan 84] R. J. Sullivan, "Look Ahead, Never Back," *Infrastructure—Maintenance and Repair of Public Works, Annals of the New York Academy of Sciences,* Vol. 431, Dec. 1984, pp. 15–17.

3

Planning, Needs Assessment, and Performance Indicators

3.1 Infrastructure Planning

Planning is a focus word for dealing with future-related activities that are concerned with achieving desired goals. As shown in Figure 1.9, planning functions are primarily concentrated at the systemwide or network level and deal with financing, budgeting, and policy issues. Modern concepts of planning recognize two distinct approaches [Binder 92]:

- *Strategic planning* is generally long-range and reflects financial and business aspects of planning, involving senior administration and/or corporate managers.

- *Tactical planning* usually reflects technical aspects of facilities and involves technical managers who are responsible for facility management and future expansion within the bounds of the strategic plans in consultation with senior administration and/or corporate managers.

Major emphasis of tactical planning is on network-level needs, including such activities as preparing and updating master plans, assessing needs and budgets, predicting future demands and developing strategies to preserve and upgrade facilities, and annual and multiyear work programming.

A practical and effective infrastructure management system (IMS) integrates planning, design and construction with the service-life activities of maintenance, rehabilitation and renovation, replacement and reconstruction (M,R&R). This concept is illustrated in Figures 1.8 and 2.3. Cost-effective M,R&R management must be integrated with planning, design, and construction.

3.1.1 Modeling and simulation

Physical facilities are subjected to many conditions, depending upon their location and type of loading or usage. The complexity of the behavior and performance of a facility is further increased due to construction quality problems, material degradation and aging, and their interrelationships. Because of these factors, a coordinated framework of systems analysis concepts is appropriate to infrastructure management. This idea was first conceptualized by Hudson et al. for pavement design [Hudson 68]. A simplified diagram of this modeling process is shown in Figure 3.1.

The scientific and engineering aspects of a systems problem span a spectrum of activities [Haas 94]:

- Use of physical observation or measurements to characterize behavior.
- Statements of mathematical models that describe or approximate the physical phenomena.
- Development of a system for prescribed behavior using the mathematical models.
- Physical realization of the system.

Therefore, it is essential to formulate a system in terms of physical or mathematical models or to use computer simulation to develop the

Figure 3.1 A simplified diagram of the major steps of modeling and system analysis to improve project-level pavement activities.

necessary outputs. As an example, consider the bridge infrastructure in California. The project-level IMS, in this case bridge management system (BMS), is concerned with improved bridge design that is resistant to seismic loading. An ideal way to look at selected alternate designs is with computer simulation.

The computer models involve earthquake-like motions based on previously recorded earthquakes in the region. Subsequently, the model outputs acceleration, stresses, and deformation, which are compared with the measured observations, wherever possible. Once a design and the simulation results are verified, then further simulations can be done under different design inputs and magnitudes of seismic motion to optimize a specific design. This modeling process also allows studying of the use of new, innovative, and high-performance materials of construction.

Network-level IMS (BMS in this example) is concerned with setting priorities for M,R&R work on substandard bridges in the network that are posted, near failure, approaching a critical condition threshold, and/or needing strengthening in the active seismic areas in California.

3.1.2 Space use planning

For planning of new facilities and M,R&R actions, it is essential to examine the space use and forecast the usage demand on the facility for the selected performance period during the service life of the facility. This will assist in selecting the most appropriate combination of materials and design to achieve satisfactory performance. Adequate consideration should be given to space use and future demand on the facility during the planning stage.

The concept of a three-dimensional space for planning of infrastructure is useful for maximizing the available space in modern urban areas. Land-use planning has been traditionally used by architects, engineers, and planners. This is a two-dimensional view of the planning needs. Most of the master plans and zoning regulations using this two-dimensional approach are prepared for cities, counties, and regional metropolitan agencies. The concept of "air space" has been restricted to the aviation facilities and airways. Similarly, underground and subsurface space has been solely used for subways, sewer and water lines, utilities, and shipping channels. "Space use" planning [Bragdon 95] is a more accurate approach for today's crowded urban regions, as illustrated in Figure 3.2.

Citing many examples of space use planning in Japan, Europe, and Washington State, Bragdon (1995) outlines a strategic master planning process considering and utilizing vertical and horizontal dimensions. Examples include multilevel golf courses in Tokyo, commer-

Figure 3.2   Three-dimensional space planning. [after Bragdon 95].

cial/office complexes in Seattle Freeway Park over Interstate 5 in Washington, and the historic bridge on the Po River in Florence, Italy, with mixed recreational, transportation, and commercial uses. The three-dimensional space concept and other innovative aviation and intermodal transportation concepts are being pursued at the recently established National Aviation and Transportation (NAT) Center on Long Island, New York, and at the NAFTA (North American Free Trade Agreement) Intermodal Transportation Institute at the University of Texas at El Paso [Bragdon 96]. This discussion points out that future construction and M,R&R work should consider the air and underground spaces in addition to the existing ground structures, to generate cost-effective alternatives.

3.1.3   Demand forecasting

The accurate prediction of future demand on an infrastructure facility is crucial to the selection of an appropriate alternative for new construction or M,R&R actions. Many statistical and analytical tools are available for developing forecasting models. These tools include cross-correlation analysis, regional market-share methods, regression analysis, time-series models, and neural network methods.

All of these techniques and models require historical data of possible explanatory variables and demand (response) variables. If the his-

Figure 3.3   Typical trend variation of data and possible model forms.

torical data is not available, then either the data for a similar situation/location or simulated data must be used to develop preliminary demand forecasting models. Plotting of historical data can show the possible form of models if age or usage indicator (e.g., cumulative traffic on a bridge) is used as a single explanatory variable (independent variable). Figure 3.3 illustrates some examples of possible linear and nonlinear forms of predictive models.

Model development is a three-step process:

1. Preliminary model based on historical data
2. Model verification using another set of data
3. Model calibration using alternate data collected under different conditions

Note that some nonlinear models can be linearized by a transformation of variables, such as using the logarithm of the variable, as shown in the bottom right plot. These modeling techniques are further discussed in Chapters 4 and 8.

Once a predictive equation is verified and calibrated, it can be used to predict the demand in a future year by plugging in the known or estimated values of the independent variable(s) for that year.

3.1.4 Environmental impact studies

For capital investment projects, environmental impact studies are often required during the planning stage to comply with state and federal regulations. The effect of planned development must be considered on surrounding communities, water bodies, wetlands, ecological systems, air quality, surface and subsurface contaminations, noise pollution, and other areas of community concern. The National Environmental Policy act of 1969 (Public Law 91-190) requires a detailed statement on the environmental impact of the proposed action and effects on the quality of the human environment [FAA 86]. There are four broad groups of factors that must be considered to evaluate the impact of any infrastructure development: (1) pollution; (2) ecological; (3) social; and (4) engineering. Some examples of these factors are:

- *Pollution factors:* air quality, water quality, noise, construction impacts on surface and subsurface soil contamination, waste-water treatment, appropriate waste disposal

- *Ecological factors:* wetlands, coastal zones, wildlife and waterfowl, endangered species, animal and bird habitats, flora and fauna, landscape and drainage, ecosystem disturbance

- *Social factors:* displacement and relocation of residences and businesses, parkland and recreational areas, historical and archeological sites, cultural and religious places, natural and scenic beauty, land development

- *Engineering factors:* storm-water drainage, flood hazards, use of energy and natural resources, costs and benefits of alternatives

Demand forecasts are also used for environmental impact analysis and effects of the development on the above factors and related areas of community concerns. For examples, noise pollution and their effects on surrounding communities are studied whenever an airport or highway expansion and construction is planned.

3.1.5 Safety and security

Safety and security are important considerations for planning, designing, and operating infrastructure facilities. Security is needed to: (1) prevent losses caused by theft, vandalism, and arson; (2) minimize the risk of possible safety hazards to the occupants and users of the facility; and (3) enforce measures for complying with the applicable laws to avoid liability claims. Planning for safety and security will depend upon risk assessment; for example, a low-risk or a high-risk facility. The most important federal laws pertaining to safe and healthful working and operating conditions was passed in 1970 and led to the creation of the Occupational Safety and Health Administration (OSHA).

3.1.6 ADA concerns

The Americans with Disabilities Act (ADA), passed in 1990, is one of the most significant federal regulations to guarantee access and physical accommodations to people with disabilities at work and public places. The law is applied to all organizations with 15 or more employees. The implementation of ADA provisions requires awareness and sensitivity to the psychological and physical environment and removal of architectural and communications barriers. All existing facilities and new facilities must accommodate the ADA requirements. Future planning must, therefore, take ADA seriously, as suggested by Prior in the following key points of influence of ADA [after Prior 94]:

- ADA has changed how people think about themselves in relationship to the workplace.
- ADA is a reflection of heightened human concerns.
- ADA is making us aware that we can do a better job when it comes to workplace accommodations.
- ADA should be viewed as a blueprint for the workplace of the future to help standardize workplace and worker expectations.

3.2 Examples of Planning Studies

Planning studies involve population projections, land-use and space-use predictions, usage (traffic) demand projections, and economic studies. The following case studies are presented to illustrate the application of analytical tools for facility planning.

3.2.1 An airport planning study example

The air-travel market is sensitive to prevailing business cycles, and it requires frequent updating of travel-demand forecasts. Airline passenger data collected at the Robert Mueller Municipal Airport in Austin, Texas, were used for a planning study in 1983. The following regression equations developed in the study suggest sales-tax revenue as a strong predictor of annual airline passenger data, as shown in Figure 3.4 [Uddin 84].

Figure 3.4 Observed and estimated annual airline passengers at Austin Municipal Airport. [Uddin 84].

$$PAX = 2071959.8 + 0.1809 \, (STR) - 6.2428 \, (POP) \quad R^2 = 0.987 \quad (3.1)$$

$$PAX = 0.1081 \, (STR) \qquad\qquad\qquad\qquad R^2 = 0.991 \quad (3.2)$$

$$\log_n(PAX) = 249.79123 - 466193.63 \, (1/YEAR) \quad R^2 = 0.960 \quad (3.3)$$

where *PAX* = total yearly airline passengers,
     STR = annual sales tax revenue in dollars, and
     POP = population of metropolitan area (thousands).

To project passenger forecasts using eqs. (3.1) and (3.2), the independent variable sales tax revenue (*STR*) also should be projected for the future year of choice. This is not always possible and it needs to be estimated first. However, the last equation is easy to use, because it contains only *YEAR* as an independent variable. In this study, the equations are extrapolated to predict travel demands in future years. This has some possible dangers and should be used with discretion.

The annual arriving and departing passengers were 2.5 million in 1983, as compared to the eq. (3.3) forecast of 2.4 million and 2.39 million predicted by a Box-Jenkins time-series model. Monthly time-series data was used to develop the Box-Jenkins ARIMA (autoregressive integrated moving average) model that can predict reliable seasonal variation, as shown in Figure 3.5.

The study also investigated the impact of these projections on aviation (runway, taxiway, and gates, etc.) and ground (terminal building,

Figure 3.5 Plots of the observed and estimated series of ARIMA model. [Uddin 84].

parking areas, gates, etc.) facilities, noise pollution due to projected increase in flight operations, and air-space problems due to the proximity of Bergstrom Air force Base [Uddin 84]. Planning of alternative sites was actively initiated to meet the future needs because the existing site had limited available space at the time of that study.

The airport has witnessed a steady growth in recent decades. Bergstrom Air Force Base was closed in 1991, selected by the City of Austin as the prime choice for the site of a new airport, and approved in 1991 by the Federal Aviation Administration (FAA) as Austin-Bergstrom International Airport [Amick 96].

3.2.2 A transportation planning study example

Montgomery County, Maryland, a large municipality north of Washington, D.C., has experienced rapid growth in employment and housing during the past several decades. The county's economic base is largely focused on information and communications technologies, biotechnologies, activities of the United States Government, and support services. In the late 1980s, the county had 350,000 jobs; 700,000 people living in 270,000 households; and one-fourth of the work force employed in the District of Columbia. The county completed a comprehensive growth policy study in 1989 for the future 30 years to assess choices for transportation [Replogle 90].

The study considered the following four land-use scenarios: *fast* but balanced growth, *slow* but balanced growth, *jobs* favoring employ-

Figure 3.6 Montgomery County origin mode share for home-to-work trips. [Replogle 90].

ment growth, and *housing* favoring housing growth. These were tested against the following mobility choices:

- Auto—Continue current policies and build out the highway master plan.
- Van—Add a network of high-occupancy vehicle (HOV) lanes.
- Rail—Add to Auto a light-rail network with certain assumptions to increase the parking fee and road pricing to effectively double the cost of automobile operation.

Figure 3.6 shows the mode shares for selected land-use and mobility choices based on analysis using the logit-mode choice model and future population and travel-demand projections. The Rail choice allows the county to meet its traffic-congestion standards, depending on the land-use balance between housing and jobs [Replogle 90]. The study findings point to the need for significant changes in the direction of master-plan development and infrastructure planning for future.

3.3 Life-cycle Management

In the postconstruction stage of infrastructure management, in-service evaluation of the infrastructure should be given high priority. Appropriate rules and checklists should be established regarding the use of the facility and maintenance/repair. Unfortunately, deferred maintenance has been the general rule in most public infrastructure including public buildings. For example, the average age of buildings at the University of Mississippi Campus in Oxford, Mississippi, is 40 years, with four buildings over 100 years old [Miss 94]. Many old buildings need extensive repair and renovation to comply with current building codes and ADA requirements. According to an article in *Building Operating Management* magazine (February 1990), by 1991 the amount of deferred maintenance costs in public school facilities was $14 billion, and college and universities had over $60 billion in maintenance, renovation, and new construction that had been deferred [Binder 92].

In reality, satisfactory performance and service quality cannot be preserved very long unless a life-cycle management plan is put into operation, preferably when the facility is opened for use. The management plan should consider the following activities:

- Rules for appropriate use of the facility
- Regulations for routine/minor maintenance due to normal use and aging
- Plans for emergency management of fire, accident, natural disaster (tornado, floods, earthquake, etc.), or arsenal sabotage
- Program of scheduled maintenance of the equipment and structure
- Framework and methodology for planning condition and demand-responsive maintenance, rehabilitation, and renovation as well as replacement or reconstruction (M,R&R) actions; framework should also include analysis of "do-nothing" and deferred maintenance actions
- Financial management plan to pay for the operation and life-cycle M,R&R requirements

3.4 Infrastructure Service Life

The most important component of life-cycle analysis is the estimate of service life of a facility. Infrastructure service life depends on design and construction methods, usage and environment, and in-service maintenance and operation practices. Service life for specific examples of any class of infrastructure may still vary greatly. Service life is not the same as design life or economic life. The terminology used in this book is described in the following sections.

3.4.1 Terminology related to service-life needs

The following terminology is based on Building Research Board publication *Pay Now or Pay Later* [BRB 91]:

Service Life: "The period in years over which a building, component, or subsystem provides adequate performance; a technical parameter that depends on design, construction quality, operations and maintenance practices, use, and environmental factors; different from economic life." This definition is equally applicable to all categories of infrastructure assets.

Performance: "The degree to which a building or other facility serves its users and fulfills the purpose for which it was built or acquired." In other words, performance is history of serviceability that shows the quality and length of service that a facility provides to its users.

3.4.2 Evaluation of infrastructure service life

Service life is the period in years from the time of completion of the facility to the time when the complete facility or its components are expected to reach a state where it cannot provide acceptable service because of physical deterioration, poor performance, functional obsolescence, or unacceptably high operating costs. Evaluation of the service life of infrastructure assets is quite complex, because different components of a facility can have varying ranges of possible service life. Service life of critical structural components should be taken as a representative estimate to plan new construction, major repair and reconstruction. At best, only an estimate of an average service life can be made based on: (1) an acceptable level of performance; and (2) average life of the facilities in each group or similar infrastructure facility after falling below an acceptable limit or performance or at failure.

In general, service life of a public infrastructure or privately owned building is not less than 40 years. Service life can be estimated from the historical infrastructure database using survivor techniques. Systems engineering approach is particularly useful to improve the performance model and predict service life based on some minimum condition acceptance criteria.

3.4.2.1 Survivor curve method.
The survivor curve shows the number of units of a property (miles of water pipelines, bridges, original cost, or percentage of units, etc.) that survive in service at given ages. Property surviving is generally expressed in percentage of the base cost at zero age. Figure 3.7 shows a survivor curve developed for the box-culvert data [Winfrey 69]. The area under the survivor curve is a direct measure of the average service life of the property units. The probable life of the surviving units at any age can be calculated from the remaining area by dividing the remaining area by the amount surviving of that age. The top of the probable life curve is the maximum number of years expected for failure in the database. A line driven vertically down from the top of the probable life curve intersects the survivor curve and gives the expected average life of the structure. Figure 3.8 shows a survivor curve for pavement rehabilitation

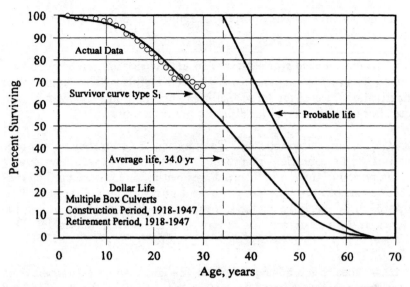

Figure 3.7   A survivor curve for box culverts based on the dollar units. [Winfrey 69].

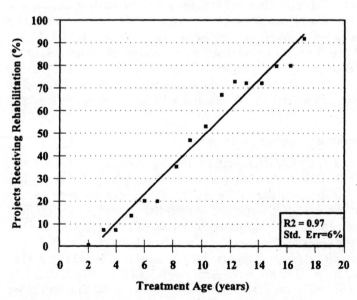

Figure 3.8   A survivor curve for SAMs applied on state routes in Arizona. [Flintsch 94].

treatment using stress-absorbing membranes (SAM) for state routes in Arizona [Flintsch 94].

3.4.2.2 Reference to previous experience. Service life can be estimated based on previous experience with such facilities. This approach is particularly useful for large facilities such as dams, nuclear plants, etc., but may be in error for different environments.

3.4.2.3 Performance modeling. The physical deterioration rate can be estimated by condition monitoring and in-service evaluation over a short period of time, and future deterioration and failure predicted as a function of age, load/demand, and environmental factors. This approach is specially useful to predict service life of major components and structures made of new and innovative materials. This subject will be discussed in-depth in Chapter 8.

3.4.2.4 Accelerated testing. Service life can also be estimated from accelerated tests designed to subject the facility to the demand/load until failure is achieved in a short time. The service life can then be assessed from the interpolation and/or careful extrapolation of the data for the facility in the same region or other environmental and general condition. Durability and service life for new construction materials is often assessed by accelerated laboratory testing. Probably the AASHO (American Association of State Highway Officials) Road Test, constructed and tested to failure within a few years during 1957–62, has been the best source of performance data and prediction of service life of asphalt and concrete roads [HRB 62].

3.4.3 Example estimates of service life

Service life is generally expressed in number of years. In cases where service life is difficult to estimate accurately, it is still useful when items are ranked in order of durability of the primary construction materials and assigned expected service life based on experience. Service life can also be expressed in terms of load repetitions (e.g., total number of aircraft coverage for an airport runway or total number of equivalent single-axle truckloads in the life cycle of a road pavement), or in other cases, as periods/cycles of use (e.g., total kilowatt hours or total cumulative running time in hours). The prevailing load, environmental, ground, and operating conditions should always be considered to predict the service life. These factors may also interact with the material of construction and the constructed facility as a whole. These interactions may lead to a shorter service life for the facility than its individual components. For example, a building wall

TABLE 3.1   General Service Lives for Highway Components [Winfrey 69]

Highway Components	Years
Right-of-way land	75 to 100
Right-of-way damages (suggested write-off period)	10 to 30
Right-of-way buildings to be moved or destroyed (suggested write-off period)	10 to 30
Earthwork	60 to 100
Culvert and small drainage facilities	25 to 50
Retaining walls and general concrete work	40 to 75
Riprap and other bank protection	20 to 50
Bridge and other major structures	50 to 75
Granular roadway surfaces	3 to 10
Low-type bituminous surfaces	12 to 20
Rigid and flexible high-type pavements	18 to 30
Signs and traffic-control devices	5 to 20

TABLE 3.2   General Service Lives of Buildings, and Comparison of Different Countries

Building Facility	Canada [CSA 94]	UK [BS 92]	Japan [AIJ 93]	
Industrial Buildings	25–49	Minimum 30	25–40 or more	Minimum 25
Commercial, Health, Education, Residential	50–99	Minimum 60	60–100 or more	Minimum 60
Civic, Monumental, National Heritage	Minimum 100	Minimum 120	60–100 or more	Minimum 60

or a bridge structure made of bricks may have a lower service life than the predicted service life of bricks.

Table 3.1 lists general ranges of service lives for highway components [Winfrey 69]. Table 3.2 shows a comparison of building lives based on codes and recommendations extracted from relevant document sources in Japan, Canada, and England. Overall service life of some infrastructure facilities is recommended in Table 8.1 in Chapter 8.

3.5   Infrastructure Needs Assessment

One key to effective infrastructure management is the establishment of network needs assessment on a regular basis and timely planning of M,R&R actions. Historically, this has been done using the past experience and judgement of the facility and maintenance engineers. This ad-hoc approach is less effective because of the expansion of infrastructure assets, changing demands on systems, and the growing shortage of funds for construction and maintenance actions at all levels of government.

The Urban Institute's six-volume report series *Guides to Managing Urban Capital* was an early comprehensive study of capital planning and budgeting practices of 40 local government agencies. The study recommends three basic strategies that a government can employ to reduce capital investment and facility maintenance problems [Urban 84]:

Strategy 1   Better identify capital needs and priorities, to screen out marginal needs and to make the best use of available funds.

Strategy 2   Build community support for facility maintenance and repair and reinvestment.

Strategy 3   Find new revenue sources, or reorganize the local revenue system so that it provides a stable source of revenues to maintain and replace basic facilities.

Most agencies operate under constrained budget conditions with competitions from other important items of public spending; therefore, a rational methodology of infrastructure needs assessment is required to convince management and legislators to allocate available funds fairly. The establishment of needs does not by itself indicate what M,R&R alternative should be considered and which is most cost-effective. The M,R&R selection and priority programming methods will be discussed in Chapters 11, 13, and 15.

3.6   Infrastructure Performance

Performance indicator of infrastructure must evaluate the quality of service provided by the facility for needs assessment. Functional performance of a facility is usually assessed from the user perspective. Hatry lists a number of performance measures including the following: quality of service and system effectiveness to meet the expectations of the users, productivity and efficiency, and resource utilization and cost-effectiveness [Urban 84].

3.6.1   Performance indicators

Performance indicators can be grouped into four broad categories:

- Service and User Perception
- Safety and Sufficiency
- Physical Condition

- Structural Integrity/Load-Carrying Capacity

3.6.2 Examples of performance indicators

All infrastructure facilities include several structural and nonstructural components. Performance of the individual components may differ with the material and usage of the component. It is also desirable to establish an overall performance indicator, with weight given to the structural integrity and service to users. Examples of performance indicators for various different infrastructure groups are designated below.

3.6.2.1 Transportation infrastructure.

Highways, roads, and streets constitute the backbone of transportation infrastructure. Pavement structural thickness and surface smoothness requirements vary according to its intended use; however, there are common performance indicators based on pavement-condition assessment; for example, a structural-capacity index based on deflection response and a distress index based on distress type, severity, and extent [Haas 94]. Such indexes are generally graduated from 100 to 0, where 100 represent the best possible condition and 0 represents the worst or failed condition. A third performance indicator is based on surface-riding quality. The objective measurement of surface-ride quality is performed by measuring surface smoothness or roughness. Smoothness and roughness represent the opposite ends of the same scale. A composite performance indicator is based on a combination of two or more of the pavement condition attributes, using appropriate weights. Other components of paved facilities (excluding bridges) are markings, signs, and roadside appurtenances or furniture items. These are subjected to environmental forces and occasional failures because of accidents. Their performance can be conveniently evaluated using an overall condition rating on a scale of 10 or 100 (the best) and 0 (the worst or failed) that is linked with the maintenance requirement.

In the case of a railroad facility, the track provides the primary load-carrying component and riding surface. Condition deterioration of a track segment will depend upon the condition of ballast, track gradient and alignment, ties and joints, wheel distribution of load and annual traffic, train speed, and environmental parameters. An overall condition index from 100 (the best) to 0 (failed and unusable) can be established based on the critical geometry, ballast depth, load, and environmental parameters. A track quality index (TQI) has been formulated that considers different track structural parameters (bad-tie counts, deflection under load), load and speed, environmental factors, and maintenance quality [Fazio 80].

Bridges are important components of all types of ground-transportation facilities. Bridges may be a part of an interchange, a grade separation, or a crossing over a water body or a river. Each major component of a bridge (deck, superstructure, substructure, and foundation) will perform differently and, therefore, requires a different condition rating. The national bridge inventory program combines these different ratings into a composite overall sufficiency rating (SR) on a scale of 100 (the best condition) to 0 (the worst condition) and uses it to identify deficient bridges [Hudson 87].

3.6.2.2 Water and waste-water infrastructure. The condition deterioration of a water supply and distribution system is primarily dependent on the material of the main pipeline network, joints, subsurface moisture and soil, demand on the system (for example, gallons of water pumped annually), infiltration of ground water, blockage to flow, and environmental effects on service life. The same is true for main sewer lines. However, from the user's and from the owner agency's points of view, number of breaks per year, number of failures per 1000 km, and leakage of water per year are the most important performance indicators.

3.6.2.3 Waste management infrastructure. Waste water from pipelines goes to treatment plants. Other types of waste collection and treatment sites are landfills. Facility-specific performance indicators are needed in such cases. These facilities do not have direct user impacts until there is a serious breakdown or fire incident that causes pollutants to spread and threaten the health and safety of public in the vicinity. Leakage quantity per year or number of breakdowns per year are possible performance indicators.

References

[AIJ 93] Architectural Institute of Japan, *The English Edition of Principal Guide for Service Life Planning of Buildings*, Japan, 1993.

[Amick 96] J. P. Amick, and J. Almond, "Reconstruction of Bergstrom Air Force Base to Austin-Bergstrom International Airport," *Rebuilding Inner City Airports, Proceedings of the 24th International Air Transportation Conference*, American Society of Civil Engineers, Louisville, Kentucky, June 5–7, 1996, pp. 1–11.

[Binder 92] S. Binder, *Strategic Corporate Facilities Management*, McGraw-Hill, Inc., New York, 1992.

[Bragdon 95] C. R. Bragdon, "Intermodal Transportation Planning for the 21st Century, a New Paradigm: An Urgent Call for Action," presented at the 74th Annual Meeting of the Transportation Research Board, National Research Council, Washington, D.C., 1995.

[Bragdon 96] C. Bragdon and C. Berkowitz, "Solving Aviation and Intermodal Transportation Related Issue, a New Prototype for the 21st Century," *Rebuilding Inner City Airports, Proceedings of the 24th International Air Transportation Conference*, American Society of Civil Engineers, Louisville, KY., June 5–7, 1996, pp. 212–222.

[BRB 91]   Building Research Board, *Pay Now or Pay Later,* National Research Council, Washington, D.C., 1991.

[BS 92]   Guide to Durability of Buildings and Building Elements, Products and Components, BS 7543:1992, British Standards Institution, London, 1992.

[CSA 94]   Guideline on Durability in Buildings, CSA S478, Draft 9, Canadian Standards Association, Sept. 1994.

[FAA 86]   Federal Aviation Administration, "Policies and Procedures for Considering Environmental Impacts," *Order 1050.1,* Washington, D.C., 1986.

[Fazio 80]   A. E. Fazio, and R. Prybella, "Development of an Analytical Approach to Track Maintenance Planning," in *TRR 744,* Transportation Research Board, National Research Council, Washington, D.C., 1980.

[Flintsch 94]   G. W. Flintsch, L. A. Scofield, and J. P. Zaniewski, "Network-Level Performance Evaluation of Asphalt-Rubber Pavement Treatments in Arizona," in *TRR 1435,* Transportation Research Board, National Research Council, Washington, D.C., 1994.

[Haas 94]   R. Haas, W. R. Hudson, and J. P. Zaniewski, *Modern Pavement Management,* Krieger Publishing Company, Malabar, Fla., 1994.

[HRB 62]   Highway Research Board, "The AASHO Road Test, Report 5—Pavement Research," *Special Report 61-E,* National Research Council, Washington, D.C., 1962.

[Hudson 68]   W. R. Hudson, F. N. Finn, B. F. McCullough, K. Nair, and B. A. Vallerga, "Systems Approach to Pavement Systems Formulation, Performance Definition, and Materials Characterization," *Final Report NCHRP Project 1-10,* Materials Research and Development Inc., March 1968.

[Hudson 87]   S. W. Hudson, R. F. Carmichael III, L. O. Moser, W. R. Hudson, and W. J. Wilkes, "Bridge Management Systems," *NCHRP Report 300,* National Cooperative Highway Research Program, Transportation Research Board, National Research Council, Washington, D.C., 1987.

[Miss 94]   "University Buildings Continue to Deteriorate," *The Daily Mississippian,* Vol. 87, No. 28, Sept. 29, 1994.

[Prior 94]   A. P. Prior, "ADA and the New Work Place Culture," *Civil Engineering News,* March 1994, pp. 25–26.

[Replogle 90]   M. Replogle, "Computer Transportation Models for Land Use Regulations and Master Planning in Montgomery County, Maryland," in *TRR 1262,* Transportation Research Board, National Research Council, Washington, D.C., 1990, pp. 91–100.

[Uddin 84]   W. Uddin, B. F. McCullough, and M. B. Crawford, "Methodology for Forecasting Air Travel and Airport Expansion Needs," in *TRR 1025,* Transportation Research Board, National Research Council, Washington, D.C., 1984.

[Urban 84]   The Urban Institute, *Guides to Managing Urban Capital,* vol. I, H. P. Harty and G. E. Peterson (eds.), Washington, D.C., 1984.

[Winfrey 69]   R. Winfrey, *Economic Analysis for Highways,* International Textbook Company, 1969.

Information Management and Decision Support Systems

Database Management, Data Needs, Analysis

4.1 Overview of Information Management

Information support and management is a critical step for the effective and successful operation of any infrastructure-management system. Access to information that is both correct and timely is essential for the satisfactory conduct of the management process. This applies equally to all key players in the organization, including planning, design, construction, and maintenance departments.

4.1.1 Information technologies

During the last decade, computer-based information technologies have become broadly and deeply integrated with management. The rapid development of electronic database storage and retrieval capabilities is the primary force behind this development. The increased popularity of personal computers and the evolution of software have also played an important role. In the early 1970s, database management system (DBMS) software, for example, was considered exotic and unreliable, whereas it is now regarded as stable and mundane [Begley 95]. Begley and Sturrock identify key information technologies for material science and engineering [Begley 95], such as relational and object-oriented database systems, expert systems, and multimedia, that either have or very likely will significantly impact the development of many applications in the areas of IMS. Other information technologies like video logging, neural networks, case-based reasoning, and virtual reality also offer potential for future infrastructure-management applications.

4.1.2 Decision support systems

Generally, the term *decision support system* (DSS) refers to the use of computers to store, analyze, and display information that is used to support decision-making. DSS implies, however, more than just data-processing business as usual; it includes analysis models that generate results useful for making rational decisions. In other words, DSS organizes the processing, analysis, and delivery of information necessary for decision making. The use of an information support system, database management, and analytical studies can help engineers make better decisions through: (1) improved identification and information of the infrastructure assets; (2) access to condition data, usage, and history; (3) delineation of problem areas; (4) methodologies for needs assessment; (5) evaluation of alternative solutions; (6) projection of work programs and budgets; and (7) priority setting of projects and programs of inspection and evaluation schedules. Therefore, DSS is an integral component of IMS.

DSS is not drastically different from the traditional staff activity of giving recommendations to the boss for making appropriate decisions. Before computers, databases existed as paper copies and were often poorly organized. Every department had its own paper files, and both information management and decision making were, if used at all, customized and more expensive than necessary, and relied heavily on personal preferences and judgment. These past practices have been replaced by computerized procedures to meet the challenge of today's decision-making management, which provide better service to the public.

Once a DSS is established, the preparation of annual and multi-year work programs for maintenance, rehabilitation/renovation, and replacement/reconstruction (M,R&R) is streamlined as an automatic DSS output. Aside from the increased productivity, plans can be continually updated by providing feedback to the database. DSS should be designed to suit the needs and resources available to an agency, and to function within the primary organizational structure. DSS is also increasingly used for operations; for example, in utility companies, plant production, transport services, and water-resource management. Figure 4.1 shows an example of a DSS framework of a pavement management system (PMS) for roadway infrastructure [Uddin 95], and a DSS example for a bridge-management system (BMS) is illustrated in Figure 4.2 [Hudson 87].

Successful implementation of a DSS must consider the flow of requests for decision support from the decision maker to the technical support staff. The two main activities in a DSS are data management and study of alternatives. These activities generate meaningful results or "knowledge" from data or "information" to support decision making.

Figure 4.1   A DSS framework for road infrastructure. [Uddin 95].

The primary role of a DSS is thus to use data, together with the necessary analytical models, to produce the decision-support rationales.

A central database is the heart of a DSS. The database typically consists of many database files. A computerized database system has several advantages over paper-based records keeping [FHWA 90], including:

- Data are stored in a compact space and shared by all users.
- Storage and retrieval of the data are much faster than a manual method, permitting the data to be updated on a regular basis and facilitating the use of the information.
- Use and processing of data are centrally controlled.

A centralized database has several advantages to the agency. However, the advantages of centralization cannot be realized without

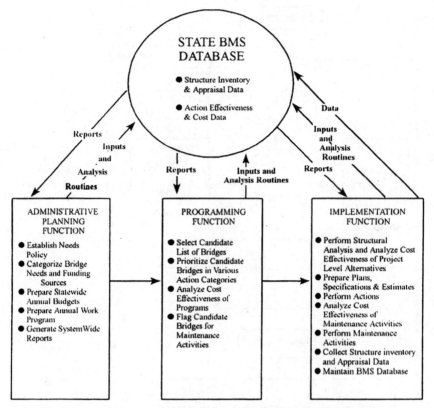

Figure 4.2   A DSS proposed for bridge infrastructure. [after Hudson 87].

a properly designed and maintained system. Advantages of a centralized system are [FHWA 90, Haas 94]:

- Redundancy is reduced by having each piece of data stored in only one place. This also avoids inconsistencies when updating the data files.

- Data can be shared for various applications. Frequently, the data needed for pavement management is collected by different divisions within the agency. Having a centralized database ensures that all divisions will have access to the needed information.

- Standards can be enforced in terms of data formats, naming, and documentation.

- Security restrictions can be applied to control the flow of the data and the updating of the data.

- Data integrity can be maintained by controlling the database updating and using integrity checking whenever an update is made.

- Conflicting requirements of the individual users can be balanced to optimize the database for the agency. This is particularly important when sharing data between divisions.

An important consideration in the development of a database management approach is the temporal and spatial identification of the data. Temporal identification is accomplished by storing the data by reference of time and/or date. It is important to establish historical records of construction, maintenance, and evaluation data. Spatial identification requires being able to physically relate the data to the location of a facility in the infrastructure network. Spatial referencing is accomplished through the section definition process, in which homogeneous sections or components with similar characteristics are identified and their physical boundaries and descriptions are established. Geocoding (geographical coordinate description) and the use of a geographical information system (GIS) can enhance this process.

Public works agencies have been using traditional decision practices since the evolution of urban society in the early 20th century in the United States. Planners in most cities are now either trying to improve the existing systems or are seeking a new approach for managing their infrastructure assets and associated services. GIS technology has emerged as a useful tool for developing a comprehensive management system so that all of the municipal infrastructures, such as pavement, bridges, water supply, waste water, sewer, gas, and electricity, can be integrated on a common platform to improve managerial decisions. This concept of an integrated infrastructure management system is illustrated in Figure 4.3, in which GIS is the common location reference system [Zhang 94].

4.2 Database Development and Management

4.2.1 System design

The term *data base* written as two separate words, or *database* written as a single word, refers to a large collection of data in a computer, organized in such a way that it can be expanded, updated, and retrieved rapidly for various uses. The term *database*, written as a single word and adopted throughout this book, also means a specific group of data within the structure of a database-management software system. The database may be organized as a single file or as multiple files or sets. The need for database software packages should be identified at the outset of system design, based on the information

Figure 4.3 A concept of an integrated overall infrastructure management system. [Zhang 94].

processing, analysis methodologies, and the scope of other specific requirements of decision support systems. Figure 4.4 shows the inter-relationships of computer hardware and software.

Software serves as the interface among users, hardware, and data. It allows the user to access data by giving instructions to the hardware. There are three important components for database systems: (1) operating system; (2) database management software; and (3) application programs. Generally, operating systems and database management software are commercially available off-the-shelf, while application programs are usually unique to a particular application.

4.2.2 Operating systems

The operating system is the first-level software required by the computer hardware. No hardware will work without an operating system. Therefore, the operating system will typically be preselected by the

HARDWARE

Figure 4.4 Interrelationships of computer software and hardware.
[after FHWA 90].

data-processing section when the computer is purchased, rather than
being a choice to make when setting up a database. Microsoft DOS is
a typical operating system for microcomputers, UNIX for minis and
workstations, and MVS-XA for mainframes. Because of the ease of
graphical user interfaces, Windows-based operating systems, such as
Microsoft's Windows 3.1, Windows 95, and Windows NT, are becoming
more popular. Apple's Macintosh operating system is another popular
Windows operating system.

Microcomputers that rely on 16-bit processors and operating system, such as DOS and Windows 3.1, are less expensive and easy to
maintain. They have become common platforms for many mapping
and database-management software packages. Many users, however,
demand capabilities that exceed the limits of microcomputers. The
capabilities of microcomputers can be enhanced by the addition of
special processing boards, larger random-access memory (RAM), and
mass storage devices [Antenucci 91]. Despite the development of
faster microcomputer processors, such as Intel's Pentium processors,
larger hard-disk space, and better graphics resolution, microcomputers are somewhat limited for handling large volumes of data and complex analysis requirements. While DOS- and Windows-based micro-

computers will continue to have a strong presence in the single-user market, the popularity of 32-bit workstations is expected to increase rapidly for the multiuser market among the local, state, regional, and national agencies responsible for managing infrastructure assets.

The workstations, employing UNIX or other operating systems optimized for 32-bit processors, deliver great processing speed and can support sophisticated software, high-resolution graphics, large RAM, and mass storage. Following the current trend of computer hardware technology, it seems apparent that decreasing prices in future years will make these workstation systems more attractive to organizations that formerly would have purchased a relatively cheaper microcomputer system. Workstations also permit the establishment of better local-area networks (LAN) for multi-user environments to share data and output devices, as compared to the LAN capability offered by microcomputers.

Currently most of the large-scale databases (in excess of 50 gigabytes) are established on mainframes and supercomputers. These large computers are preferred over other computers because of their massive storage capacity. The architecture of these large computers allows multiple mass-storage disk packs. The disk packs can be easily daisy-chained together. Each of these disk packs can have 300 or more gigabytes. Workstations can retrieve data from the mainframe databases by connecting mainframes on the wide-area network connections.

4.2.3   Database-management software

Database-management software is the next level of software above the operating system. This software platform allows the user to define data structures and models without worrying how the data are physically stored on the hardware. Database-management software provides a mechanism for an application programmer to write programs that can perform various data-access, retrieve, and manipulation functions. They also provide the capability for certain queries, such as *ad hoc* queries, to be made on the database. Three kinds of database management software are available: (1) relational, (2) hierarchical, and (3) network. The relational database management software (RDBMS) is the most recent and by far the most used. The other two types are used primarily in the mainframe environment. Hierarchical databases are structured by users as a tree in which users must start at the root and follow specific branches to get to a particular "leaf" they want. Network databases are similar to hierarchical ones, except any particular "leaf" can be attached to more than one branch. Relational databases (for example DB2 for IBM mainframes and OS/2 workstations, or dBase IV for microcomputers) allow users to organize the database as a collection of tables that can be linked together.

4.2.4 Application programs

Application programs are software specifically written to accomplish data-manipulation and analysis functions, and to provide input interfaces and design output generators, as required by the end-users. Inventory and condition-assessment computer programs for a railroad company or for the water department of a city public-works agency are examples of application programs. High-level programming languages have been adopted by the data-processing community as a means to develop and modify applications more quickly. Fourth-generation languages, usually associated with a particular database-management package, have greatly increased the efficiency and productivity of programmers in application development work. Some GIS software packages also provide an integrated high-level programming language. In the future, fifth-generation languages using less structured syntax, sometimes referred to as "natural languages," will be a favorable addition to geographic information technology [Antenucci 91]. Manufacturers of GIS packages rely to a greater degree on third-party relational database packages. Many vendors have started offering the option of integrating or linking their software to a variety of relational packages.

4.2.5 Database management and data manipulation

Since the early 1980s, the data-processing community as a whole has embraced the relational model and relational techniques. Recent advances in relational database software have greatly enhanced the user interface, and the architecture of computer processors has become more suitable for relational data structures. The current database-management software packages combine the flexibility of relational models with an ability to perform interactive transaction processing on large databases. Advances in high-level-query and programming languages have prompted the popularity of relational database systems. Data-query languages have adapted English-like commands to perform complex retrievals from the database. A recently accepted standard is the structured query language (SQL). Initially developed by IBM, SQL is used by many relational database software and GIS software developers.

In relational database software, the size of infrastructure networks and data activities influence the memory and hard-disk storage space requirements. Typically, nongraphic relational databases can be managed using relatively small (less than one megabyte) hard-disk space. Memory and storage-space requirements for graphic databases can be ten times higher than for the nongraphic databases. Users in many organizations require 2 to 5 gigabytes, and the trend suggests tenfold increases in demand during the course of the next decade [Antenucci 91]. For large-size agencies the use of large mainframes, supercomputers, and workstations on a network becomes a necessity, particularly when the database is expected to fill up 50 or more gigabytes.

4.2.6 Geographic and nongeographic data

GIS software (available for workstations and microcomputers) provides the user with mapping capabilities, and links attribute databases to several map features. This is useful for the end-user to visualize the physical layout and specific attributes of an infrastructure facility. Figure 4.5 illustrates this concept. A GIS database usually consists of two basic types of data: geographic and nongeographic. Each type has specific characteristics and requirements for efficient data storage, processing, and display. The following discussion is summarized from

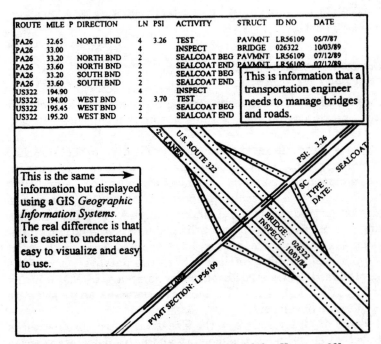

Figure 4.5   An example of a visual display of data. [after Keystone 93].

the work of Antenucci et al. The database of a GIS typically is composed of multiple sets of geographic and nongeographic data managed by the database engine of the GIS. The data in a GIS may include digital coding of map features, logical geographical relationships among features, and nongeographic data that describe characteristics of the corresponding features [Antenucci 91].

4.2.6.1 Nongeographic data. Nongeographic data are representations of the characteristics or attributes of map features. They are stored in conventional numerical formats, although data such as graphic images can be linked to nongeographic data with GIS technology. The term *nongeographic* is used here to differentiate those data that do not represent the geographic features. They are related to geographic locations through common data fields or identifiers. In GIS software, nongeographic data are managed separately from the geographic data due to their different characteristics.

4.2.6.2 Geographic data. Geographic data represent map features in a computer-readable form. Geographic data use six types of graphic elements (i.e., points, lines, areas, grid, cells, pixels, and symbols) to depict map features and annotation. Just like the AUTOCAD software for computer-aided graphics, a series of layers is often used to describe the graphic component of the GIS database, each of which contains map features that are related functionally. Figure 4.6 shows the layering concept. Each layer is a set of homogeneous features that is registered positional to the other database layers through the common coordinate system. The separation into layers is based on logical relationships. One major purpose of the layering is to simplify the combination of features for display. This electronic layering scheme is comparable to a series of overlays in a manual mapping system.

4.2.7 Geocoding

Geocoding (geographic coordinates) data are important for graphic displays. Figure 4.7 shows commonly used coordinate systems. Typically, data can be obtained by using a digitizer on a paper map, a differential GPS (global positioning system) receiver, satellite imagery data, the Topological Integrated Geographic Encoding and Referencing (TIGER) files, and digitized data files produced from computer-aided drafting graphics software, such as AUTOCAD. The TIGER system was developed by the U.S. Geological Survey (USGS) and the Bureau of the Census for the 1990 census of the United States. The TIGER system is the first comprehensive digital map of

Soils

Administrative/
Political Districts

Zoning

Utilities

Parcels

Topographic
Contours

Planimetric
Features

Geodetic/Survey
Control

Figure 4.6 An Illustration of the database layering concept. [after Antenucci 91].

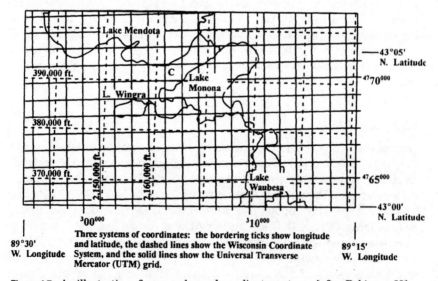

Figure 4.7 An illustration of commonly used coordinate systems. [after Robinson 69].

the nation's roads and streets at a scale of 1:100,000, street addresses in urban areas, railroads, and all significant hydrographic features [Antenucci 91].

4.3 Data Needs

4.3.1 Data requirements for infrastructure management

Databases and data-management software are needed in any department of an agency that is responsible for infrastructure management. Some of these databases may be used by one or more departments. For example, financial databases are usually used by accounting departments; contract and project information data are used by contract and construction departments; and maintenance work databases are used by accounting and maintenance departments. The key to effective database management in IMS is the linkage among various databases so that the necessary information is easily accessible. However, the following data are essential to a generic IMS:

- Inventory data for physical descriptions of infrastructure facilities, including construction and metrical data, preferably geographically based
- Usage-history data (for example, water usage for water-distribution systems, aircraft coverages for airport pavement)
- Condition monitoring and evaluation data
- Real-time monitoring data
- Maintenance history and operation data
- Maintenance intervention criteria, decision criteria, maintenance policy, and unit cost and budget data
- Design and analysis data
- Maintenance and construction priority list data

4.3.2 Data details

Classification of data needs and levels of detail are very important issues because data collection can be a very expensive effort. Haas et al. proposed a three-level concept of information detail and complexity of models, as shown in Figure 2.4 [after Haas 94]. The lower-left and upper-right triangles are infeasible areas because while detailed data is required for analysis at the project-level, detail and complexity of models restrict the amount of detail of data at the network. Similarly, performance modeling for project-level design applications requires a detailed material property database, as described by Uddin for road-pavement applications [Uddin 95].

A study by the World Bank recognizes five functional levels of data needs [Paterson 90], which offer a framework to establish infrastructure management data needs for specific agency objectives:

Sectoral-Level: Aggregation of data from the road infrastructure system; e.g., annual statistics and road-user charges for comparison with the education sector, for example.

Network-Level: Planning, programming, and budgeting: strategic planning (three to five years medium-term needs); strategic regional transportation planning; tactical network-level work programming for M,R&R

Project-Level: Project-level programming for selection and extent of treatment and exact location; may require intensive data collection and analysis.

Operational-Level: Facility and operation management related to construction, maintenance, traffic, and safety.

Research and Development: Data needs are more detailed and precise than that for project level and operational level. Data are usually study-specific.

Element-specific data detail is then best defined for groups of similar information, since these groups have features in common for more than one application and form a natural basis for collection and storage in a database. For each category, the specific data needs relevant to a functional application can be identified. The amount of detail required for these various applications can be seen to increase progressively from the overall summary statistics through planning, programming, design, and research. In accordance with Figure 2.4, as the system becomes more complex the data-collection efforts increase per unit from network-level analysis to project-level analysis. More sophisticated methods would be required to gather the more detailed data required for each group of information. This leads to the definition of four information-quality levels that represent the ranges of both the data requirements and the methodologies for collecting the data.

The road inventory and pavement condition data needs can be identified for four information quality levels (IQLs), suggested in the World Bank's report [Paterson 90]. These IQLs are listed in Table 4.1. Level 1 is the greatest level of detail and is typically benchmark-type information that will be collected only on a project-level basis. Level II is the greatest amount of detail that would be gathered at a network level and usually requires sampling. Level III detail is adequate

TABLE 4.1   Classification of Information Quality Level (IQL) and Detail, with Example Application to Road Infrastructure [after Paterson 90]

IQL	Amount of Detail
I	Most comprehensive level of detail; normally used at high-class project-level design and research studies; requires high level of institutional resources to collect and process the data
II	Level of detail sufficient for programming models and standard design methods; would usually require sampling and automated acquisition equipment for network-level programming; requires reliable institutional sources
III	Level of detail sufficient for planning and standard programming models for full network coverage; network data collection can be combined using automated and manual methods
IV	Level suitable for the simplest planning and programming models; suitable for standardized road-design catalogs; not adequate for high-class project-level design; the simplest data-collection methods, either entirely manual or partly semi-automated; requires the least resources

for planning purposes, and data may be collected using a combination of manual and automated methods. Level IV is the minimum detail needed for generating the basic statistics and other information on the network for planning models. The concept of IQL from roads also can be applied to other infrastructure-management applications.

The general classification of detail in terms of the number of parameters is complicated by the question of reliability. It is possible, for example, to compensate for some lack of detail by increasing the reliability of each data point by raising the accuracy of measurement and/or intensifying the measurement sample rate.

4.3.3   Data terminology and data dictionary

The agency should compile a comprehensive data file of all pertinent local terminology and definitions for various data elements (inventory, condition, equipment, data-summary statistics, and indices) and appropriate units of measurement. The adoption of international standards for new data elements would be useful. This is crucial for consistency among agency staff and other users and should be considered as a step to ensure quality.

4.4   Analysis and Modeling Techniques

Data analysis is an integral part of IMS that distinguishes IMS from a traditional information-management system. Analysis requires the use of models and analytical techniques. Models serve various purposes, including: (1) a performance model to predict life; (2) a model to backcalculate material properties; (3) a failure-analysis model, such as fatigue models; or (4) cost models, such as maintenance-cost or user-cost models.

Nishijima outlines several steps in material property modeling [Nishijima 95]:

- Catch general trend and scatter of the data
- Evaluate quality of the data set
- Deduce characteristic values of the property
- Compare the data with other data (statistical tests)
- Predict values by interpolation and extrapolation
- Determine confidence limits of the property
- Obtain mathematical expressions for further tests

These steps are also generally valid for other types of data and model development. A brief overview of available modeling techniques is presented next. These techniques require a reliable database as the first step.

4.4.1 Regression analysis

Traditionally, data have been analyzed using statistical regression analysis. This method is commonly used to develop empirical models by estimating parameters and coefficients of independent or explanatory variables in mathematical relationships that can explain most of the variations in the dependent or predictor variable [Box 78]. A scattergram of the data should be reviewed first, to estimate the model shape. Analysis of variance (ANOVA) should precede regression analysis to identify statistically significant independent variables. The regression model may be linear or nonlinear. Regression modeling has its limitations, especially if the scattergram does not show a known model shape; if many different independent variables influence the dependent variable; or if the goodness of fit is low (low R-square values). Many new analysis and modeling techniques have emerged during the last decade.

4.4.2 Expert systems

The definition of an expert system is "interactive software that uses simplified methods to drastically limit search time in large problems for certain tasks." The simplest architectural model of an expert system, shown in Figure 4.8, indicates that the main components of expert systems are a knowledge base and an inference engine. Expert systems differ from conventional computer programs in two ways:

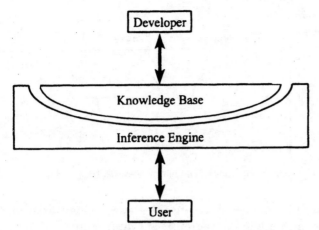

Figure 4.8 Architecture of an expert system. [after Begley 95].

1. Expert systems work with domain-specific knowledge that is symbolic as well as numerical.

2. Expert systems use domain-specific methods that are heuristic (simplified) as well as algorithmic.

4.4.3 Object-oriented database management system (OODBMS)

Object-oriented database management systems (OODBMS) employ data structures and programming concepts originally developed in object-oriented programming (OOP) languages. OOP languages were developed to address cost and efficiency issues associated with the implementation of large software applications. Object-oriented software uses cell-like components called *objects* that communicate with one another via messages, much as cells communicate through chemical messages in the human body. An object, or abstract data type, is a block of code that combines data with operations to manipulate that data. The development of OODBMS solved the database query problems associated with CAD applications in automotive and aerospace industries. Object-oriented databases can be viewed as database-management systems based on relationships and functional characteristics of data where data access paths are built into the object types [Begley 95].

4.4.4 Artificial neural networks

Artificial neural networks (ANNs) are computing systems made up of several simple, highly interconnected elements that process information by dynamic response to external inputs. Neural network archi-

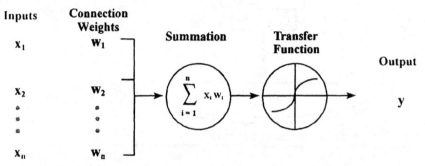

Figure 4.9 The processing elements in an artificial neural network. [Begley 95].

tecture was inspired by studies of brains [Begley 95, Ghaboussi 92]. The basic model of an individual neuron is shown in Figure 4.9 for a simple artificial neural network. It does not execute a series of fixed instructions like a traditional computer program or statistical analysis; rather, it responds, in parallel, to the inputs presented to it during a training period. Each neuron computes and maintains a dynamic variable called activation, S_j, which is a function of the sum of inputs that arrive via weighted pathways. The input from a particular pathway is an incoming signal, X_i, multiplied by the weight w_{ij}, of that pathway. The outgoing signal from neuron i is the activity of the unit, computed as $O_j = f(S_j)$ where $f(S_j)$ is called the activation or transfer function, usually a binary threshold or a sigmoid function. The neuron can also have a bias term, b_j, which acts as a form of threshold. The relationship between incoming signals and the outgoing signal can be expressed as:

$$S_j = \Sigma W_{ij} X_i + b_j \qquad (4.1a)$$

$$O_j = f(S_j) \qquad (4.1b)$$

$$f(S_j) = 1/[1 + e^{-S_j}] \qquad (4.1c)$$

The major variables of a neural network are network topology (the number of nodes and their connectivity), rules of computation of the activations of the processing units, rules of propagation, and the rules of self-organization and learning [Ghaboussi 92]. One of the common types of rules involves back-propagation networks, also referred to as *feed-forward networks*. The processing unit in back-propagation neural networks are arranged in layers. Each neural network has an input layer, an output layer, and a number of hidden layers. A neural network performs "computations" by propagating changes in activation between the processors. Propagation takes place in a feed-forward manner, from the input layer to the output layer. Processing ele-

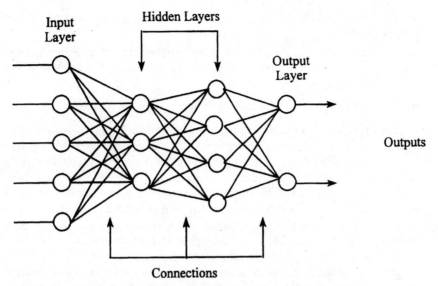

Figure 4.10 ANN processing elements and interconnections. [Begley 95].

ments and interconnections from a sample neural network are shown in Figure 4.10. Each unit receives input from all the units in the proceeding layer and sends its output to all the units in the next layer.

The neural network gains its knowledge through training. A supervised training method is commonly used to train feed-forward networks. In supervised learning, a set of training data (consisting of input/output patterns) is presented to the network, one example at a time. For each set, the input pattern is propagated through the network and the resulting output pattern is compared to the target. A learning algorithm is employed to incrementally adjust the connection weights to reduce the difference between the calculated and target outputs. This is done by back-propagating the error through the network. Once trained, the network will provide an approximate functional mapping of any input pattern onto its corresponding output pattern. Once a neural network has processed all of its training data and achieved a state of equilibrium, new input data can be presented for evaluation.

4.5 Database Security

When working with high-volume IMS databases, it is imperative to develop and maintain a comprehensive data backup and security system. Such a system should be based upon:

- Daily and weekly backup, with the provision for rotation of the backup medium
- Permanent backup at appropriate intervals; at a minimum it should be done on an annual basis, with the provision for long-term rotation of the backup medium
- Off-site storage of permanent backup

Considerable time and money is invested in the development of databases, and without the implementation of appropriate security protocols the agency could face serious consequences. These protocols should be documented, the responsible staff trained, and frequent surveillance undertaken to ensure continued implementation. Off-site storage of the latest permanent backup is essential to recover the lost data and other information in case of a catastrophe like fire, flood, earthquake, or arson. The protocols should include standard formats for backup identification, storage, and retrieval.

A variety of backup media are available, depending on the computer hardware and operating software. These range from floppy diskettes, 8-mm tapes, and spools to removable high-memory optical disks. In all options some type of data-compression utility is used to compress, or zip, the data files to maximize the capacity of storage medium.

An excellent example of the ultimate benefit of offsite backup storage is the digital GIS database of the road infrastructure of Kuwait. During the 1990–1991 Persian Gulf war, all computerized digital road network databases were reportedly destroyed or corrupted; however, a backup copy available outside the country was used by Intergraph to print a detailed road map of Kuwait [Intergraph 91]. Another example of total devastation is the Oklahoma City federal building that was completely destroyed by a terrorist bombing on April 19, 1995 [Time 95]. These incidences emphasize the need to keep an off-site permanent backup copy of IMS databases.

4.6   Data Quality Control and Quality Assurance Issues

Consistency of data collection and processing procedures and database accuracy should be given top priority. Principles of quality management should be followed to achieve quality outputs/products [Hayden 89, Uddin 91a].

4.6.1   Quality control (QC)

- *Definition:* The operational techniques and activities that are used to fulfill requirements for quality. In other words, it is the process of doing and checking the work before releasing it.

- Quality control is exercised through detailed step-by-step office and field procedures, instructions, and operational guidelines contained in the appropriate manuals. Examples: road pavement inventory manual [Uddin 91b]; pavement distress manuals [Shahin 90, SHRP 93, Uddin 91c].

- Productivity level for all IMS data-handling and database operations should be checked, and a system outlining the minimum desired level of productivity should be based upon assigned equipment and staff resources.

- QC principles and tailored procedures should be used continually for all future IMS operations.

4.6.2 Quality assurance (QA)

- *Definition:* All planned and systematic actions necessary to provide reasonable confidence that a product or service will satisfy quality requirements.

- For effectiveness, QA usually requires continuing verification, inspection, and evaluation of the operations and outputs. This can be accomplished for IMS operations through on-the-job training, manual instructions, dedicated data collection, and processing forms and procedures.

- IMS activities must be appropriately documented for records and follow-up.

- Work supervisor(s) should ensure data consistency and implement quality assurance through a system of routing files, follow-up, and random checks.

- QA principles and specially developed protocols should be continually used for all future IMS operations.

References

[Antenucci 91]   J. C. Antenucci, K. Brown, P. L. Croswell, M. J. Kevany, and H. Archer, *Geographic Information Systems,* Chapman & Hall, New York, 1991.

[Begley 95]   E. F. Begley, and C. P. Sturrock, " Matching Information Technologies with the Objectives of Materials Data Users," *Computerization and Networking of Materials Databases,* Vol. 4: *ASTM STP 1257,* American Society for Testing and Materials, Philadelphia, Pa., 1995, pp. 253–280.

[Box 78]   G. E. P. Box, W. G. Hunter, and J. S. Hunter, *Statistics for Experimenters: An Introduction to Design, Data Analysis, and Model Building,* John Wiley and Sons, New York, 1978.

[FHWA 90]   "An Advanced Course in Pavement Management," course text, Federal Highway Administration, U.S. Department of Transportation, Washington, D.C., 1990.

[Ghaboussi 92]   J. Ghaboussi, "Potential Application of Neuro-Biological Computational Models in Geotechnical Engineering," *Numerical Models in Geomechanics,* Pande and Pietruszczak (editors), Rotterdam, 1992.

[Haas 94]   R. Haas, W. R. Hudson, and J. P. Zaniewski, *Modern Pavement Management,* Krieger Publishing Company, Malabar, Fla., 1994.

[Hayden 89]   W. M. Hayden, "The Effective A/E Quality Management Program: How To Do It," short course notebook, American Society of Civil Engineers, New York, 1989.

[Hudson 87]   S. W. Hudson, R. F. Carmichael III, L. O. Moser, W. R. Hudson, and W. J. Wilkes, "Bridge Management Systems," *NCHRP Report 300,* Transportation Research Board, National Research Council, Washington, D.C., 1987.

[Intergraph 91]   W. Uddin, Personal Communications to Intergraph, Reston, Va., Oct. 1991.

[Keystone 93]   *Geographic Information System,* product brochure, Keystone Management Systems Inc., 1993.

[Nishijima 95]   S. Nishijima, "Common Data Processing Needs for Material Databases," *Computerization and Networking of Materials Databases:* Fourth Volume, ASTM STP 1257, American Society for Testing and Materials, Philadelphia, 1995, pp. 9–19.

[Paterson 90]   W. D. O. Paterson, and T. Scullion, "Information Systems for Road Management: Draft Guidelines on System Design and Data Issues," *Technical Paper INU77,* Infrastructure and Urban Development Department, The World Bank, Washington, D.C., 1990.

[Robinson 69]   A. N. Robinson and R. D. Sale, *Elements of Cartography,* John Wiley & Sons, 1969.

[Shahin 90]   M. Y. Shahin and J. A. Walter, "Pavement Maintenance Management for Roads and Streets Using the PAVER System," *USACERL Technical Report M-90/05,* Champaign, Ill., 1990.

[SHRP 93]   "Distress Identification Manual for the Long-Term Pavement Performance Project," *Report SHRP-P-338,* Strategic Highway Research Program, National Research Council, Washington, D.C., 1993.

[Time 95]   "Oklahoma Bombing," *Time,* April 1995.

[Uddin 91a]   W. Uddin, "Dubai Road Pavement Management System (DRPMS)— DRPMS Manual of Data Processing, Analysis & Generation of DRPMS Reports," *U.N. Expert's Report No. PM-14, Dubai Municipality, U.N. Project UAE/85/012,* United Nations Centre for Human Settlements (Habitat), Dubai, United Arab Emirates, July 1991a.

[Uddin 91b]   W. Uddin, "Dubai Road Pavement Management System (DRPMS)— DRPMS Manual of Road Network Partitioning and Inventory Data Collection," *U.N. Expert's Report No. PM-12, Dubai Municipality, U.N. Project UAE/85/012,* United Nations Centre for Human Settlements (Habitat), Dubai, United Arab Emirates, July 1991b.

[Uddin 91c]   W. Uddin, "Dubai Road Pavement Management System (DRPMS)— DRPMS Manual of Pavement Inspection and Distress Data Collection," *U.N. Expert's Report No. PM-9, Dubai Municipality, U.N. Project UAE/85/012,* United Nations Centre for Human Settlements (Habitat), Dubai, United Arab Emirates, July 1991c.

[Uddin 95]   W. Uddin, "Pavement Material Property Databases for Pavement Management Applications," *Computerization and Networking of Materials Databases,* Vol. 4: *ASTM STP 1257,* American Society for Testing and Materials, Philadelphia, Pa., 1995, pp. 96–109.

[Zhang 94]   Z. Zhang, T. Dossey, J. Weissmann, and W. R. Hudson, "GIS Integrated Pavement and Infrastructure Management in Urban Areas," in *Transportation Research Record 1429,* Transportation Research Board, National Research Council, Washington, D.C., 1994, pp. 84–89.

Inventory, Historical and Environmental Data

5.1 Infrastructure Management Data Needs

5.1.1 Background

The implementation of IMS in the form of pavement-management systems (PMS) since 1970 [Haas 94, Hudson 94], and in bridge-management systems (BMS) since the late 1980s [Hudson 87, Golabi 92] clearly demonstrate the requirement for valid inventory or identification databases. The national bridge inspection standard (NBIS) database serves for BMS, and each state individually has covered their needs in PMS. These systems demonstrate the importance of good data and a well-balanced database in managing the design, construction, and M,R&R (maintenance, rehabilitation/renovation, and replacement/reconstruction) activities for infrastructure facilities. The database for any IMS involves the following:

- Good identification and location-referencing data
- Inventory of physical description and construction historical data of past actions
- Condition monitoring and evaluation data
- Usage data and maintenance history
- Feedback system to update and improve IMS activities

These system components are essential for any IMS, as well as for an integrated IMS that ties together several types of infrastructure. In some agencies, the location-referencing and physical inventory data attributes for infrastructure are available, but are not systemati-

TABLE 5.1   Model Interactions Among the Transportation Information
Systems. [after Hudson 94].

Transportation infrastructure	Bridge	Pavement	Safety	Public transportation	Congestion	Intermodal
Bridge	—	Major	Major	Major	Minor	Major
Pavement	Major	—	Minor	Major	Major	Minor
Safety	Major	Minor	—	Major	Major	Major
Public transportation	Major	Major	Minor	—	Major	Minor
Congestion	Minor	Major	Minor	Major	—	Minor
Intermodal	Major	Minor	Minor	Minor	Minor	—
Emissions	Minor	Major	Minor	Minor	Major	Minor

cally coordinated. Recently, some agencies have made progress in linking their primary databases, as reported for New York [Cohn 95] and California [Robison 96]. These initial efforts may be time-consuming, but lead to the advantages of common computer-systems operation and quick access to the primary inventory database for emergency management. In the case of New York City, a $1.7 million integrated system of computer-aided drafting and design (CADD) and GIS provides a single electronic information channel for everything from improving pothole repair and road design to developing tree-planting plans and mapping the city's bridges. The City of Oakland, California, collected over $6 million in compensation from the Federal Emergency Management Agency (FEMA) for *post facto* damage to city streets after a devastating fire in October 1991 on the basis of a detailed pavement-management report that quantified the damage to streets resulting from heavy reconstruction traffic.

Typical GIS software accesses database files, and a GIS database file can be easily linked with large IMS databases. Figure 4.3 in Chapter 4 illustrates an integrated database based on GIS. Such an integrated database is cost-effective for facilities that share common information needs. This is illustrated by Table 5.1, which models interactions among the transportation subsystems [after Hudson 94].

5.1.2 Data details and data collection frequency

A proper database will contain details ranging from a few mandatory elements to a long list of data elements, depending on the objectives of the IMS. The use of specific data will vary with the functional level of IMS (i.e., network-level planning function or project-level design). As discussed in Chapter 2 (Figure 2.4), the data details and system complexity required increases from the network-level to project-level IMS functions. An IMS decision support system (DSS) database consists of a number of files that depend on the data-collection frequency and updating. The database files, listed in Table 5.2, are section-specific, facility-specific, and applicable to all sections in the network. The files include the following data types:

- Section-specific data (inventory, usage, and condition evaluation)
- Facility-specific data (M,R&R policies and unit costs, performance models)

TABLE 5.2   The Interrelationship of IMS Data Collection Frequency and Database Update

DSS component	Database Update		Database file
	Mandatory	Desirable	
1. Inventory (location reference, physical description, construction type and data, functional class, M,R&R history)	Once (original construction)	After every major M,R&R and capital improvement	Section-specific
2. Usage* (historical usage record and future estimate)	For a new IMS, back cast for best estimates, then annually	Monthly and seasonal in addition to yearly total	Section-specific
3. Condition evaluation* (condition monitoring and evaluation data)	Once every 1–3 years (on sampling basis)	More than once every 1–5 years (on 100% network)	Section-specific
4. M,R&R Policies and costs* (short-term, annual, long-term)	Once (used for analysis)	Current data (annual update of costs)	Facility-specific
5. Performance prediction (condition deterioration, effect of M,R&R on condition)	For life-cycle analysis	Every year	Facility-specific
6. Economic analysis (analysis period, economic evaluation parameters)	Once (used for analysis)	Multiple interest and inflation rates	Network-specific (global)
7. M,R&R Program reports* (economic evaluation, short-term and long-term M,R&R work program reports)	Every year annual M,R&R program	3 years, 5 years, 10 years Multi-year M,R&R program	Network-specific with facility-specific outputs
8. Environment* (rainfall, temperature, etc.)	Annually	Annually	Use NOAA or statewide access

*Not inventory data. See Chapter 6 for detailed discussion of item 3 (condition evaluation data). See Chapter 15 for discussion of item 7 (M,R&R program reports).

- Network-specific data (economic analysis parameters, M,R&R work programs)

Data updating is a key activity that applies not only to usage data and condition evaluation data, but also to changes in the inventory resulting from major M,R&R projects. Items 2, 3, 4, 7, and 8 in Table 5.2 are not inventory but are shown here for completeness.

5.2   Aspects of Inventory Data

An inventory of infrastructure facilities generally refers to the data related to the physical features, including structural components, physical dimensions, material properties, and construction details (including date of original construction). For existing facilities, it is essential to determine and record in the database a history of all construction, rehabilitation, capital improvement, and other physical changes in the structure, drainage, and so forth. The M,R&R work and cost history, as well as the historical record of usage, are also important data attributes. Most characteristics do not change from year to year, but any items that change because of M,R&R should be updated immediately. The database design requires that all data be defined with respect to specific section and location-referencing codes. Some of the inventory data attributes are mandatory for M,R&R analysis to generate network-level M,R&R programs. Other data elements may be desirable for a broad and detailed M,R&R analysis and for project-level analysis of alternative design strategies. The following major classes of inventory and historical data are important to the IMS databases:

- Section identification and location
- Functional class
- Geometrical data
- Structural data
- Material type and property data
- Appurtenance data
- Construction and M,R&R history
- Costs data
- Environmental data
- Usage history

5.2.1   Network partitioning

One of the important initial steps for building an inventory database is the partitioning of the network into homogeneous management units or sections. Each of these management units is given a unique identification number that is the backbone organization of the IMS inventory, historical, and condition databases and all related analysis and managements reports.

Generally, an IMS unit or section should have the same material type, geometric parameters, construction history, functional classification, operating volume, and characteristics along its entire length. Other practical aspects include easy identification on the network map and in the field, and manageable length for monitoring. For example, a road network section length will be generally 2 km (1.2 mi) or less. Local streets may have one section per block. On the other hand, an expressway would be divided into a number of sections. Some facilities stand alone like a bridge or a dedicated building structure (for example, a coliseum, monument, water tower, or air-traffic control tower) and should be treated as a unique unit.

Partitioning criteria should be established after a thorough study of the infrastructure network and appropriate numbering methodologies. The following common parameters can be used as partitioning criteria:

- Agency's jurisdiction and administrative zone boundaries

- Functional and use classifications (for example, storm sewer or sanitary sewer)

- Primary construction material and geometry/structure characteristics

- Construction or major rehabilitation/renovation/replacement project number and completion date

- Boundaries based on major operating component/equipment locations (For example, the junction of two sewers can be the beginning point for the first trunk sewer.)

- Land parcel and adjacent road references (with geographical coordinates)

5.2.2 Dynamic segmentation

Network partitioning based on location identifiers and physical descriptions establishes homogeneous sections. This approach is necessary to calculate age, area, and volume of M,R&R work, and future-condition predictions; however, the uniformity of units based on behavioral attributes, such as deflection and hydraulic characteristics, may be different. A dynamic segmentation procedure can be used to establish new boundaries of units without losing other physical inventory data in the original sections [Uddin 92]. Even a large building, such as the Pentagon in metropolitan Washington, D.C., can be subdivided into manageable units based on wing, floor, and so forth.

a1) A sewer line digitized as a string consisting of 14 sections based on sewer connection.

a2) A road digitized as a string consisting of four sections, based on intersections.

Sewer

Road

MP 1.952

MP 0.000

MP 5.247

MP 8.003   MP 9.674

b1) How the string would be segmented by application of flow characteristics.

By flow discharge characteristics

b2) How the string would be segmented by application of pavement condition attribute milepoints.

POOR

FAIR

MP 2.809

MP 0.000

by PCI (Pavement Condition Index)

MP 7.413

GOOD

MP 9.674

Figure 5.1 Dynamic segmentation based on evaluation data.

Figure 5.1 shows an example in which several main storm-sewer sections are included in larger segments based on discharge flow characteristics, and the corresponding street sections are segmented by attribute mile points using the pavement-condition rating.

Dynamic segmentation can be implemented in GIS software because of its powerful query operations. For example, the IMS database can be queried to display all sections with similar attributes, or homogeneous segments can be characterized based on selected threshold values of evaluation data.

5.2.3 Location referencing methodology

Location references should be based on well-defined points at the start and end of the section, along with the geographical coordinates at both ends. The location should be easy to identify in the field as facilitated by GPS. Postal zip code and street address must be included for any building or fixed plant facility. Location references used for a road section can also be adapted for other spatially distributed facilities that are present along the road, as shown in Figure 5.2 [after Hudson 94]. Location referencing codes should not be confused with

Figure 5.2 Example of common location reference for spatially distributed infrastructure facilities. [after Hudson 94].

record sequence numbers generated by software or input by a user. A unique number for each IMS section or management unit is important for two reasons: (1) the IMS database and analysis software packages use the information to track each section; and (2) the user identifies the section by its code, both in the field and on the reports generated. The properly designed location referencing code can be used to link numerous other infrastructure components.

5.2.4 Inventory of physical assets

Considering the quantity of possible physical inventory data (functional class, geometry, structure, material type and properties, and appurtenances), it is essential to prioritize the data and/or collect the primary data elements that are required for M,R&R analyses first. Suggested mandatory inventory data elements for different infrastructure groups are shown in Table 5.3. Functional class is an important data item for identification of IMS sections.

The two primary functions of roads are *mobility* and *accessibility*. A functional class reflects the geometrical and structural design standards. Commonly used functional classes are arterial (such as freeway, expressway, highway), collector, and local. Local streets are generally designed for accessibility, whereas high-level arterial roads are predominantly designed for high-speed mobility. The classification by jurisdictional entity (for example, private, municipal, county, state, federal) is also useful to know the type of M,R&R and funding resource. It is also possible to downgrade the information quality level (IQL) of some of the data groups [Paterson 90] using the recommenda-

TABLE 5.3   Examples of Infrastructure Inventory Data Elements

Facility	Identification/ location	Construction/geometry/ material/structure*	Costs and usage history*
(1) Highways, roads, streets, parking areas	Agency, state, county, city, name and number, section number, coordinates, use, functional class	Construction number, type and date, lanes, pavement width and length, surface type, material types and thickness, shoulder, side-walk, drainage, appurte-nance, traffic control and lighting	Total/annual construction and M,R&R costs, unit costs, annual traffic and percent vehicle type
(2) Airports: pavements, buildings, aviation facilities	Agency, state, county, city, name and number, section number, coordinates, use, functional class	Construction number, type and date, airport lighting etc; see (1) for pavement data; see (8) for building data and air-traffic-control facilities	Aircraft mix, annual operations and passenger; see (1) for cost data and vehicular traffic data
(3) Bridges: interchange, over/ underpass, flyover, rail/ river crossing	Agency, state, county, city, name and number, NBIS number, coordinates, use, functional class, road	Construction number, type and date, spans, material and size of substructure and superstructure; see (1) for approach pavement and deck surface data	See (1) for cost data and vehicular traffic data
(4) Railroad: terminals, track, bridge	Agency, state, yard and terminal address, track sections and location, use, coordinates	Construction number, type and date, track geometry and material, track bed, traffic control; see (3) for bridge data; see (8) for terminal building data	Traffic mix, annu-al operations of trains/goods/ passengers; see (1) and (8) for cost data
(5) Water: treatment plants, storage, supply	Agency, state, county, city, plant number, address, tower, hydrants and pump stations, main pipeline section number, coordinates	Construction number, type and date, tower and hydrant data, soil, pipe diameter and material, pipeline lent, pipe lining, pump and valve data; see (1) for road and parking data; see (8) for plant building data	Total/annual construction and M,R&R costs, unit costs, annual water consumed, annual consumer consumption data
(6) Waste water and sewer	Agency, state, county, city, treatment plant number, address, pump stations, main pipeline section number, coordinates	Construction number, type and date, manholes, soil, pipe diameter and material, pipeline length, pipe lining, joint and pump data; see (1) for road and parking data; see (8) for plant building data	Total/annual construction and M,R&R costs, unit costs, annual sewer pumped and treated, storm sewer and con-sumer sewer data

TABLE 5.3   Examples of Infrastructure Inventory Data Elements (*Continued*)

Facility	Identification/ location	Construction/geometry/ material/structure*	Costs and usage history*
(7) Solid-waste facilities	Agency, state, county, city, treatment plant number, address, waste disposal section number, coordinates	Construction number, type and date, length and width and size of plant and fill site, equipment, soil and treatment data; see (1) for road and parking data; see (8) for plant building data	Total/annual construction and M,R&R costs, unit costs, annual waste collected and treated, street trash and consumer data
(8) Buildings: multipurpose complexes	Agency/owner, state, county, city, name, street address, code number, coordinates, use, functional class	Construction number, type and date, primary material type and data, number of floors, floor data (room number, use, length, width, height, door, window, light, air conditioning), soil and foundation data; see (1) for road and parking data	Total/annual construction and M,R&R costs, unit cost data, yearly persons served and capacity used per event, yearly revenue
(9) Power-supply facilities	Agency/owner, state, county, city, name, street address, facility number, function, coordinates, power-line section	Construction number, type and date, cable material type and data, number of poles, pole material and size, transformer stations and equipment, meters, soil; see (1) for road and parking data; see (8) for building facility	Total/annual construction and M,R&R costs, unit cost data, operation cost, yearly kilowatt and consumers served, yearly revenue

*Complete data for original construction (designated as construction number 1) and for subsequent M,R&R (designated as construction number 2 and higher). Construction number designations are used to relate all other data belonging to the particular construction project, as explained further in Sec. 5.2.5.

tion outlined in Chapter 4. For example, only pavement-surface type and thickness, collected by high-speed visual windshield survey and radar surveys, respectively, for highways represent IQL III for the structure and material data groups. This low-cost effort will suffice for network-level M,R&R planning over a relatively short time frame. Later, these data at IQL I can be collected in greater detail (material type, thicknesses, and strength of each layer using as-constructed project documents and/or coring) for overlay design at project-level.

A waste-water management system, on the other hand, consists of a number of distinct components for collection and disposal of sanitary sewage and storm waste water, where specific inventory data

Figure 5.3 Configuration of urban waste-water management system.

items can be assigned to a unit. Figure 5.3 illustrates a schematic of a typical urban waste-water management system [after Grigg 88]. The components of a comprehensive waste-water management system for sanitary sewage and storm waste water are:

- Household drains for sanitary and storm-water sewage
- Street and common-area storm-water sewage collection (gutters, ponds, channels)
- Municipal sewer collector system for sanitary and storm sewage (sewer pipes or open channel ditches, manholes, pumps)
- Main sewers (pipes and ditches)
- Treatment plants
- Sludge facilities
- Disposal of treated water

5.2.5   Construction and M,R&R history

Construction and M,R&R history should be included in the database starting with the original construction. This can be very efficiently achieved by establishing the following data pertaining to *construction number* (construction number 1 for original construction, construction number 2 and above for subsequent M,R&R work, etc.): *construction type* (original construction, M,R&R type); *construction material* (primary materials used in the original construction and M,R&R work);

construction date (date of completion of the original construction or subsequent M,R&R work); and *construction cost* (total cost of work).

Construction number is used as a key identifier to represent any major work done that can significantly affect subsequent performance. All geometric data, structural data, materials data, and cost data should be linked to the specific construction number. Construction quality and variability affect the pavement condition; therefore, data variability should be recorded in the database if at all possible. The accumulation of all historical data, collected over time, provides the basis for assessing, developing, and calibrating performance models needed for life-cycle analysis.

5.2.6 Geometry, structure, and material data

The extent of geometry, structural, and material data can vary from a few mandatory items to many hundred items, depending on the infrastructure and the intended use of the data. These data items must refer to each construction number for a given section. The mandatory geometry data include overall dimensions, length, width, and thickness of primary components. Structural and material data should include, as a minimum, material type and dimensions of the primary load-bearing component, and a list of any other components and their material type. Project-level analysis may require more detailed data for each structural component. Examples of these data for a bridge, a road section, and a building are presented in Secs. 5.6, 5.7, and 5.8. Detailed examples of the scope of needed pavement structure and material data for pavements are available in many other references [Hudson 94, Uddin 93, Uddin 95, Zhang 94].

5.2.7 Cost data

Cost history of new construction and M,R&R strategies is an important component of the inventory database. If a computerized maintenance bookkeeping procedure or maintenance management system already exists, then the past cost history can be electronically called up as needed. The cost history data can be used to develop average M,R&R unit-cost models for analyses of M,R&R strategies. In addition, database files and subfiles should be developed to store unit costs of common M,R&R actions based on the past records and bid estimates, and other agency records or commercial sources, such as the Means cost estimate catalogues [Means 93]. The unit-cost database files should be updated periodically. These are primarily used to calculate agency costs for projecting the M,R&R budget of the owner agency.

User costs are also important, including costs of travel time, vehicle operation, traffic delay due to construction, user discomfort, accidents, excess costs due to disruptions and breaks in service, user fees and taxes. For road infrastructure, very comprehensive vehicle-operating costs models have been developed by the World Bank [Chesher 87, Paterson 92], Texas Research and Development Foundation (TRDF), and the U.S. Federal Highway Administration [Zaniewski 82]. These are used in programs for network-level road-management systems, such as USER [Uddin 94]. Similarly, other examples of user costs can be cited for other types of facilities:

- *Airports:* Traveler's time cost and inconvenience due to inadequate auto parking, inefficient planning of ticket counters and gates, poor baggage-handling facilities; time cost to airlines because of delays to aircraft departures and arrivals to gates

- *Water supply:* Increased water bills to consumers because of unaccounted leakage; extra cost and inconvenience to consumers in case of loss of pressure and/or rationing of water supply

- *Electric power supply:* Consumer's lost productivity and loss to businesses due to outages and interruptions in power supply

- *Buildings:* Occupant's lost productivity and inconvenience due to inadequate lighting, inefficient air conditioning, poor air quality; revenue losses to owners of rental properties

5.2.8   Environmental data

Environmental conditions generally affect the life and performance of infrastructure facilities. Environmentally induced stress can contribute to premature fracture of concrete and metal structures. These failures occur at stress levels considerably below design stresses and may cause serious property damage and threat to human life. According to Fitzgerald, the majority of water main breaks occur where the pipe wall has weakened, e.g., corrosion of cast-iron pipes [Fitzgerald 68]. Other examples of weakening and failures induced primarily by environmental factors are interruptions in water supply and loss of water due to leakage, frost heave of building and roads, and soil contamination due to leakage in underground fuel tanks. Corrosion-induced material degradation and reduced fatigue life are serious concerns for sewer mains and water mains [O'Day 84], cooling facilities at nuclear plants, steel and other industrial plants.

The conditions for environmentally assisted stress-corrosion cracking (SCC) and corrosion-fatigue cracking (CFC) are more probable in the case of heavy usage and in aging structures. Other forms of environmentally related cracking are hydrogen stress cracking and sulfide stress cracking. These phenomena can drastically reduce the life of a structure, and reliable service-life predictions for susceptible materials are difficult to make. A common feature of each of these processes is subcritical crack growth, during which cracks grow from existing flaws or initiation locations and increase to a size at which catastrophic failure occurs [Sprowls 96]. The fatigue strength of a given material may be significantly degraded in the presence of a harsh environment, as shown in Figure 5.4 [after Sprowls 96].

Seasonal variations of temperature and water can alter soil strength and surface or subsurface drainage characteristics and can have a direct impact on structure life and performance. It can also affect the selection and cost of M,R&R strategies. While there are many techniques for ensuring adequate or good drainage, the data characterizing drainage is usually either in terms of a porosity or permeability value, or is subjectively recorded as good, fair, or poor [Haas 94].

Freeze-thaw conditions can significantly lower load-carrying capacity of pavements and break or disrupt water and sewer lines. For example, the appearance of potholes on roads and highways, a common scene in the U.S. during the spring, is affected by freeze-thaw environmental action, as shown in Figure 5.5 [Minsk 84]. Occurrence of potholes on major roads in urban America is a hot news topic each spring because of thawing and weakening of pavement layers. An

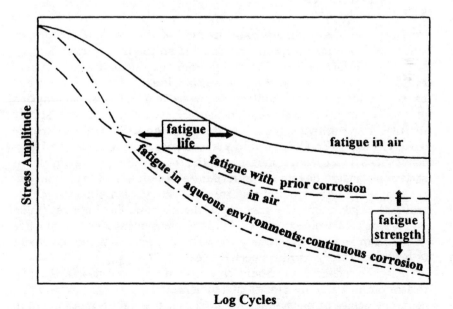

Figure 5.4 Schematic of the effects of corrosion on fatigue life and fatigue strength as shown in laboratory tests. [after Sprowls 96].

Figure 5.5 A schematic of pothole formation resulting from freeze-thaw actions. [after Minsk 84].

estimated 77,000 potholes in Baltimore and 125,000 in New York were repaired during the first four months of 1996 [CNN 96].

If the environmental conditions vary significantly across an agency's area of jurisdiction, a record of the local environmental conditions are needed to predict performance and aid in the selection of appropriate M,R&R strategies. Simple indexes of environmental conditions are often used, such as the maximum and minimum ambient temperatures, wind speed, Thornthwaite moisture index, freeze-thaw cycles, freezing index, seasonal rainfall, or an empirical "regional factor" developed by the agency.

5.2.9 Usage history

Performance of any facility is a function of its use, load history, or traffic. Information on usage and demand on the facility are required to predict performance and to assign priorities during the selection of M,R&R projects. For water and sewer facilities, the usual measure of usage is total quantity handled per day or year. Annual consumption of kilowatt-hours is a possible measure for an electric-power supply facility. For highway agencies, the average annual daily traffic (AADT), with a breakdown into percent passenger vehicles and percent trucks, is a common measure of the total traffic on the pavement section or bridge. Pavement performance modeling and bridge deterioration, on the other hand, require an estimate of the heavy vehicle traffic that generates the majority of the distress. For highway pavements, the total number of 80 kN (18 kip) equivalent single-axle loads (ESAL) can be used to estimate the vehicular load that the pavement or bridge has carried or is expected to carry.

Generally, usage data should be collected and recorded in the historical database using on-site instrumentation, such as flow meters, counters, scales at permanent weigh stations, weigh-in-motion equipment [TRB 86], and the application of appropriate demand-prediction models. Expected growth rates should also be included for such usage data. For airfield pavements, records should be maintained on the total numbers of movements of each aircraft type or class. Discharge flow of water or sewage should be recorded, if possible, on hydraulic facilities. In the case of bridges and buildings, volume usage and loading forces, such as dead and live loads, wind, and seismic forces, need to be considered.

5.3 Technologies for Inventory and Historic Data Collection

The technologies of data acquisition for infrastructure inventory items may be grouped into several classes:

- Transcription from "as-built" or "as-constructed" project records. The historical project record is the least expensive source of inventory information.

- Pedestrian observer visual survey, usually on a sampling basis. This is feasible for small-size facilities. It is also the best method to verify the information collected from other sources.

- Windshield surveys (moving observer visual survey). This method is relatively faster but yields only estimates; however, it enables greater coverage for a fixed budget.

- Photo or video logging has the advantage of permanent records. This method can be used for roads, airports, railroad tracks, subways, sewers, large water lines, and other facilities. Subsequent data reduction by either manual or automated image-processing techniques is, however, required and costly.

- Automated measurements of geometrical and structural characteristics can be done efficiently on roads, railroad tracks, and pipelines and sewers.

- Weather records can provide simple, inexpensive environmental data.

- Nondestructive testing, such as ground-penetrating radar, magnetic resonance, acoustic emission, and wave-propagation methods, are sometimes useful.

Table 5.4 summarizes several methods and options. Table 5.5 identifies methods available for field-inventory data collection on roads, parking facilities, railroads, and airports. The inventory data of physical facilities are characteristics that generally do not change from year to year, and those items that do change are updated either on the completion of a project or on an annual basis. The major data collection effort is therefore the initial one, undertaken at the time of establishing the information system for the network.

TABLE 5.4   Classifications of Inventory Data Collection Methods to Establish the Baseline History [after Paterson 90]

(1)   Transcription from "as-built" or "as-constructed" engineering records, such as plans, quality-control documents, and accounting and project records. The information on costs and M,R&R history is always obtained from the past records.

(2)* Pedestrian observer visual survey, usually on a sampling basis, with manual recording, audio or photographic recording, or electronic encoding (e.g., on a hand-held computer).

(3)* Windshield survey riding on a vehicle (moving observer visual survey) with manual recording, keyboard entry to computer-encoded memory, and audiorecording (voice-recognition) to computer memory.

(4)* Video or 35-mm photo logging, with permanent recording on photographic film (usually 35 mm), cinematic film (usually 16 mm, sometimes 8 mm), video film, or optical disc. Subsequent data reduction is performed by either manual or automated image-processing techniques.

(5)   Automated on-board measurement of geometrical and structural characteristics.

(6)   Weather records for environmental data gleaned from published data from local or national weather stations or on-line Internet sources, or site-specific instrumentation and data loggers.

(7)* Nondestructive testing, such as ground-penetrating radar, magnetic resonance, infrared imaging, acoustic emission, deflection testing, and wave-propagation methods.

*May also be used for periodic condition surveys.

TABLE 5.5   Field Encoding Methods for Inventory Data of Road and Airports [after Paterson 90]

Acquisition method	Device type	Device examples	Examples of user country
Pedestrian	Hand-held computers	PSION 2	Europe
	Hand-held computers	Husky Hunter	U.K.
Wind shield	Keyboard	Desy 2000	France
	Keyboard	ARAN[1]	Canada
	Keyboard	RST[1]	Sweden
	Graphic tablet	RIS[2]	Norway
	Voice recognition	AREV[1]	Australia
Photologging	35-mm continuous photography	GERPHO[1]	France
	35-mm continuous photography	ROADRECON[1]	Japan, U.S.
	Videologging	ARAN[1]	Canada
	Videologging	AREV[1]	Australia
	Videologging	PAVETECH[1]	U.S.

[1]Part of a multifunction dedicated vehicle
[2]"Road Inventory System"

A more detailed inventory may be required at the project level for project design. Thus, the method selected for data acquisition depends on the purpose. Historical usage data is collected by in-service monitoring on a regular periodic basis. Well-organized environmental data with periodic updates are readily available from the National Oceanic and Atmospheric Administration [NOAA 94].

5.4   Inventory Data Collection
and Processing

Once the section inventory and historical data elements are defined, the collection process begins. Depending on the sampling plan and the purpose, this task can be quick or time-consuming. Electronic data collection and transfer should be used. Agencies without a centralized database may use hard-copy records and files of field-data collection. Much of the inventory information required is in historical records, but it must be automated.

Generally, the construction history begins with the as-built plans, which provide data on the layout, dimensions, and boundaries of the facility; materials used; and year of construction. In cases where the structure or facility has evolved over time or construction records are lost, such as old bridges, water and sewer facilities, and roads, it is necessary to rely on the recollection of experienced people to estimate the construction history. In some cases when the construction history is not available, physical measurements and a survey must be used to obtain information on dimensions and structure, including nondestructive testing and coring or trenching to examine the underground structure. For about one-third of the bridges in the U.S. national bridge inventory, there is no information on foundation depth, but recently nondestructive wave propagation testing has been used to estimate this information [Aouad 96]. It is not necessary to have a separate field-investigation program to establish the construction history. The data can be collected as part of a structural evaluation over time. For example, one agency worked with the utility company using a form that work crews filled in whenever they cut a trench across a pavement to capture the thickness and material type for each layer to feedback information [Haas 94].

Data should be collected and processed systematically using formats designed for ease of use and precise recording. Data processing in the office or the field should use a formatted data sheet or template to provide direct computer entry. A notebook computer carried into the field can provide a viable method of recording data. Procedures for quality control and quality assurance are needed to ensure data accuracy. Variability and accuracy should be estimated. For example, radar estimates of layer thickness are not nearly as accurate as actual measurements. On the other hand, a precise measurement at only one point is not an adequate sample to represent the time mean value for a large facility.

5.5   Institutional Issues

Dedicated staff are important for an IMS database development plan. To fully understand the capabilities/limitations of current technology, designated IMS engineers should have prior experience in related areas and/or should attend a short course on infrastructure management. Visits to and interaction with other agencies of comparable size can offer useful insight into system development. However, this can be a blind-leading-the-blind situation if the agency visited is not competent in IMS. On-going training of assigned IMS staff should be an integral part of any IMS development plan.

5.6   Example Inventory Data System for Bridges

A bridge-inventory system differs from a linear facility, such as a road or a utility pipeline. A bridge can be considered to be one or more unique sections. It consists of several important groups of structural and functional or nonstructural components, as shown in Table 5.6 [Hudson 87]. In the structural bridge inventory, the three main data groups are deck, superstructure, and substructure. Figure 5.6 shows several major components of a typical bridge over a water crossing. The National Bridge Inspection Standards (NBIS) outlines a record of inventory and condi-

TABLE 5.6   Summary of Bridge Inventory Variables
[after Hudson 87]

Inventory data groups	Number of data elements
Identification information	13
Environment	3
Defense-importance ratings	7
Essentiality/classification/jurisdiction	20
Navigation and waterway	3
Posting information	4
Safety inventory	8
Secondary features	20
Structural inventory	14

Note: There are six M,R&R historical variables in the NBIS database.

Figure 5.6   Major components of a typical bridge.

TABLE 5.7   NBIS Code Number and Bridge Description [after FHWA 88]

Code	Description	Code	Description
00	Other	12	Arch—through
01	Slab	13	Suspension
02	Stringer/multibeam or girder	14	Stayed girder
03	Girder and floor beam system	15	Movable—lift
04	T-beam	16	Movable—bascule
05	Box beams or girders—multiple	17	Movable—swing
06	Box beams or girders—single or spread	18	Tunnel
07	Frame	19	Culvert
08	Orthotropic	20	Mixed types
09	Truss—deck	21	Channel beam
10	Truss—through	22	Channel beam
11	Arch—deck		

tion of all highway bridges and culverts with spans of 6 m (20 ft) and tunnels using a national coding guide [FHWA 88]. The most important items in the inventory are: (1) predominant material type (such as concrete, steel, timber, etc.); (2) predominant type of design/construction, as selected from Table 5.7; (3) bridge structure type; and (4) functional class. The functional classes by type of service that the bridge provides are [Xanthakos 94] highway, railroad, pedestrian, highway-railroad, waterway, highway-waterway, railroad-waterway, highway-waterway-railroad, relief for waterway, and others. Fifteen common bridge structure types have been identified in the NCHRP study on bridge-strengthening needs in the United States [Dunker 87]. These are listed in Table 5.8.

TABLE 5.8   Distribution of 15 Common Bridge Types [after Dunker 87]

NBIS item 3	Main structure type	Number of bridges	Percentage of bridges
302	Steel stringer	130,892	27.2
702	Timber stringer	58,012	12.0
101	Concrete slab	42,450	8.8
402	Continuous steel stringer	36,488	7.6
310	Steel trough truss	31,206	6.5
104	Concrete tee	26,798	5.6
502	Prestressed concrete stringer	26,654	5.5
201	Continuous concrete slab	21,958	4.6
102	Concrete stringer	16,884	3.5
505	Prestressed concrete multiple box	16,727	3.5
303	Steel girder—floor beam	9,224	1.9
204	Continuous concrete tee	7,467	1.6
111	Concrete deck arch	6,245	1.3
501	Prestressed concrete slab	5,561	1.2
504	Prestressed concrete tee	4,687	1.0
	Total	441,253	91.8

As an example of special needs, agencies in a seismic region may need to evaluate bridges for seismic retrofitting. This would require seismic rating of each bridge. The first step in the seismic rating process is an inventory to establish the following information [Buckle 87]: (1) structural characteristics, to determine the vulnerability rating; (2) seismicity of the bridge site; and (3) importance of the structure as a vital transportation link. These, as well as postearthquake evaluation, are further discussed in Chapter 7.

5.7   Example Inventory Data for a Road Section

Extensive literature has been published by the Federal Highway Administration [FHWA 89, FHWA 90], state highway agencies, and others. The Federal Aviation Administration (FAA) of the U. S. Department of Transportation [FAA 82] has implemented PMS on airport pavements. The PMS evolution and related technologies are well documented by Haas et al. [Haas 94]. Therefore, a case study of inventory database design for a road-management system for Dubai Emirate in the Persian Gulf [Uddin 91, Uddin 93] is presented for illustration. The first step was to identify the road network using the existing plans and databases and to establish network partitioning criteria, homogeneous sections, and a location-referencing methodology for Dubai.

5.7.1 Identification and historical data

The following identification and historical data are included for each section: road name, sector and community numbers (planning-zone references), road number, functional class, past construction or M,R&R project data and completion date, direction, reference chainages (start and end), centerline length, carriageway type (single or dual/divided), pavement surface type, number of through lanes in each direction, AADT, directional AADT, percent trucks, traffic count and axle-load data, and geographical coordinates. The inventory data collection form provides an adequate explanation of the data ranges and/or allowable codes for use in the office as well as in the field. Inventory databases are used immediately to generate useful summary statistics and graphs, as shown in Figure 5.7.

5.7.2 Geometric, construction, and structure data

The geometric and construction data include the key fields of section number and the following data categories: construction number and date, carriageway geometry details (for mainline pavement, median, verge, and service road sections), sidewalk/footpath data (type, length, and width), junction types and locations (for roundabout,

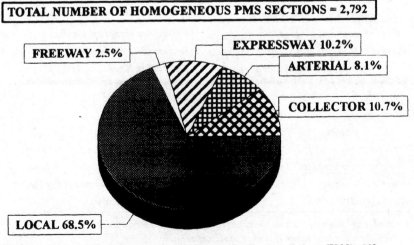

Figure 5.7 Example distribution of road sections by functional class. [Uddin 93].

intersection, T-section, and interchange), parking area (type, location, length, and width), turning-lane type (left, right, u-turn, acceleration, and deceleration) and dimension, and shoulder data (inside/outside, type, length, and width). An explanation of *construction number* is provided in Sec. 5.2.5; this data item is used to establish and record historical reference.

Construction number and date are also used to refer to pavement layer numbering and layer material description and thickness [Uddin 95]. The structural data form includes the key fields of section number and construction number. The data also includes types and dimensions of secondary structures, appurtenance, drainage, roadside safety structures, and details of pavement layer material type and thickness for carriageways and shoulders/sidewalks.

5.8 Example of Inventory Data for Buildings

A building consists of many structural and nonstructural components. The primary classification by material type includes wood-frame, masonry, concrete, and steel-frame structures [ATC 93]. The functional classification by predominant use is a long list, covering residential, commercial, industrial, public, educational, health-related, correctional, and monumental buildings. Further classification can be by design/construction type. A building life-cycle cost program, which was developed for the American Society of Testing and Materials

TABLE 5.9 Example of Inventory Data Used for Buildings at a College Campus

Campus and location data	Building specific data
Campus name and location	Facility/building name and code; GIS code
Date of evaluation and assessor's name	Access road; parking area; foundation
Facility/building name and code (Following data for each building)	Exterior and roof data and ratings (Following data for each room/space)
GIS code	Room number and name
Location on campus	Floor level
Number of floors	Room/space use[3]
Facility use[1]	Height, dimensions, and floor area
Construction material[2]	Number of doors; condition rating
Construction date	Number of windows; condition rating
Last evaluation date	Last evaluation date

[1]Administration, education, health, social, plant, laboratory, other
[2]Reinforced concrete, brick, timber, steel, other
[3]Office, elevator, library, education, rest room, storage, corridor, exterior, car park, other

(ASTM), includes a large number of maintenance and repair cost items in the categories of carpentry, electrical, plumbing, painting, air conditioning, heating, masonry, roofing, fire safety, and steam fitting [ASTM 90]. An example of an inventory data required for a general public building is shown in Table 5.9. Separate data forms for campus data and building-specific data were designed for possible use at a college campus such as the University of Mississippi at Oxford.

References

[Aouad 96]   M. F. Aouad, L. D. Olson, and F. Jalinoos, "Determination of Unknown Depth of Bridge Abutments Using the Spectral Analysis of Surface Waves (SASW) and Parallel Seismic (PS) Test Methods," *Proceedings, 2nd International Conference on Nondestructive Testing of Concrete in the Infrastructure,* Nashville, Tenn., 1996, pp. 147–153.

[ASTM 90]   "Building Maintenance, Repair, and Replacement Database (BMDB) for Life-Cycle Cost Analysis," *A User's Guide to the Computer Program,* American Society for Testing and Materials (ASTM), Philadelphia, Pa., 1990.

[ATC 93]   Applied Technology Council, "Postearthquake Safety Evaluation of Buildings Training Manual," *ATC-20-T,* funded by the Federal Emergency Management Agency, Washington, D.C., 1993.

[Buckle 87]   I. G. Buckle, R. L. Mayes, and M. R. Button, "Seismic Design and Retrofit Manual for Highway Bridges," *Report FHWA-IP-87-6,* Federal Highway Administration, McLean, Va., May 1987.

[Chesher 87]   A. Chesher, and R. Harrison, *Vehicle Operating Costs, The Highway Design and Maintenance Standards Series,* A World Bank Publication, Johns Hopkins University Press, Baltimore, Md., 1987.

[CNN 96]   *Headline News,* Cable News Network (CNN), April 27, 1996.

[Cohn 95]   F. Cohn, "New York Gets Wired," *Civil Engineering,* Vol. 65, No. 9, Sept. 1995, pp. 54–57.

[Dunker 87]   K. E. Dunker, F. W. Klaiber, and W. W. Sanders, "Bridge Strengthening Needs in the United States," in *Transportation Research Record 118,* Transportation Research Board, National Research Council, Washington, D.C., 1987.

[FAA 82]   "Guidelines and Procedures for Maintenance of Airport Pavements," *Advisory Circular AC: 150/5380-6,* Federal Aviation Administration, Washington, D.C., 1982.

[FHWA 88]   "Recording and Coding Guide to the Structure Inventory and Appraisal of the Nation's Bridges," *Report FHWA-ED-89-044,* Federal Highway Administration, U.S. Department of Transportation, Washington, D.C., 1988.

[FHWA 89]   "Pavement Management Systems, A National Perspective," *PAVEMENT Newsletter,* Federal Highway Administration, U.S. Department of Transportation, Issue 14, Spring 1989.

[FHWA 90]   "An Advanced Course in Pavement Management," course text, Federal Highway Administration, U.S. Department of Transportation, Washington, D.C., 1990.

[Fitzgerald 68]   J. H. Fitzgerald, "Corrosion as a Primary Cause of Cast Iron Main Breaks," *Journal of American Water Works Association,* Vol. 68, No. 8, 1968.

[Golabi 92]   K. Golabi, P. Thompson, and W. A. Hyman, *Pontis Technical Manual,* prepared for the Federal Highway Administration, Jan. 1992.

[Grigg 88]   N. S. Grigg, *Infrastructure Engineering and Management,* John Wiley & Sons, New York, 1988.

[Haas 94]   R. Haas, W. R. Hudson, and J. P. Zaniewski, *Modern Pavement Management,* Krieger Publishing Company, Malabar, Fla., 1994.

[Hudson 87]   S. W. Hudson, R. F. Carmichael III, L. O. Moser, W. R. Hudson, and W. J. Wilkes, "Bridge Management Systems," *NCHRP Report 300,* Transportation Research Board, National Research Council, Washington, D.C., 1987.

[Hudson 94] W. R. Hudson, and S. W. Hudson, "Pavement Management Systems Lead the Way for Infrastructure Management Systems," *Proceedings, Third International Conference on Managing Pavements,* National Research Council, Vol. 2, May 1994, pp. 99–112.

[Means 93] *Means* cost data book, 1993.

[Minsk 84] L. D. Minsk, and R. A. Eaton, "Strategies for Winter Maintenance of Pavements and Roadways," *Infrastructure—Maintenance and Repair of Public Works, Annals of the New York Academy of Sciences,* Vol. 431, Dec. 1984, pp. 155–167.

[NOAA 94] "Climatological Data," National Oceanic and Atmospheric Administration, National Climatic Center, Asheville, N.C., 1994.

[O'Day 84] D. K. O'Day, "Aging Water Supply Systems: Repair or Replace," *Infrastructure—Maintenance and Repair of Public Works, Annals of the New York Academy of Sciences,* Vol. 431, Dec. 1984, pp. 241–258.

[Paterson 90] W. D. O. Paterson and T. Scullion, "Information Systems for Road Management: Draft Guidelines on System Design and Data Issues," *Technical Paper INU77,* Infrastructure and Urban Development Department, The World Bank, Washington, D. C., 1990.

[Paterson 92] W. D. Paterson and B. Attoh-Okine, "Simplified Models of Paved Road Deterioration Based on HDM-III," in *Transportation Research Record 1344,* Transportation Research Board, National Research Council, Washington, D.C., 1992.

[Robison 96] R. Robison, "Pavement Management Pays Off," *Civil Engineering,* Vol. 66, No. 4, April 1996, pp. 44–47.

[Sprowls 96] D. O. Sprowls, "Environmental Cracking—Does It Affect You?," *ASTM Standardization News,* Vol. 24, No. 4, April 1996, pp. 24–29.

[TRB 86] J. A. Epps and C. L. Monismith, "Equipment for Obtaining Pavement Condition and Traffic Loading Data," *NCHRP Synthesis 126,* Transportation Research Board, National Research Council, Washington, D.C., 1986.

[Uddin 91] W. Uddin, "Dubai Road Pavement Management System (DRPMS)— DRPMS Manual of Road Network Partitioning & Inventory Data Collection," *U.N. Expert's Report No. PMS-12, Dubai Municipality, U.N. Project UAE/85/012,* United Nations Centre for Human Settlements (Habitat), Dubai, United Emirates, July 1991.

[Uddin 92] W. Uddin, "Highways and Urban Road Maintenance Management: Development and Operation Issues," *1992 Compendium, 6th International Pavement Management/Maintenance Exposition and Conference,* Atlanta, Ga., 1992.

[Uddin 93] W. Uddin and M. Al Tayer, " Implementation of Pavement Management Technology for Dubai Emirate Road Network," *Proceedings, 20th World Congress of International Road Federation,* Vol. IV, Madrid, Spain, May 1993, pp. 305–314.

[Uddin 94] W. Uddin, "Application of User Cost and Benefit Analysis for Pavement Management and Transportation Planning," *Proceedings, 4R Conference and Road Show,* Philadelphia, Pa., Dec. 1993, pp. 24–27.

[Uddin 95] W. Uddin, "Pavement Material Property Databases for Pavement Management Applications," *Computerization and Networking of Materials Databases,* Vol. 4, *ASTM STP 1257,* American Society for Testing and Materials, Philadelphia, Pa., 1995, pp. 96–109.

[Xanthakos 94] P. P. Xanthakos, *Theory and Design of Bridges,* John Wiley & Sons, New York, 1994.

[Zaniewski 82] J. P. Zaniewski, B. C. Butler, G. Cunningham, G. E. Elkins, M. S. Paggi, and R. Machemehl, "Vehicle Operating Costs, Field Consumption, and Pavement Type and Condition Factors," *Texas Research and Development Foundation Final Report,* Federal Highway Administration, U.S. Department of Transportation, Washington, D.C., 1982.

[Zhang 94] Z. Zhang, T. Dossey, J. Weissmann, and W. R. Hudson, "GIS Integrated Pavement and Infrastructure Management in Urban Areas," in *Transportation Research Record 1429,* Transportation Research Board, National Research Council, Washington, D.C., 1994, pp. 84–89.

6

In-Service Monitoring
and Evaluation Data

6.1 In-Service Evaluation Data Needs

In-service monitoring and evaluation of condition is an essential part of infrastructure management. Good evaluation information is required to adequately model maintenance and rehabilitation requirements and to measure the effectiveness of various maintenance and rehabilitation methods. The development and application of available and new technologies is necessary for efficient infrastructure management.

6.1.1 Monitoring and evaluation

The evaluation phase of infrastructure management involves monitoring the usage and physical condition of the infrastructure assets being managed. Monitoring involves the collection of field inspection data. Evaluation involves the analysis, interpretation, and/or judgment of the meaning of the information collected. The purpose of in-service evaluation in IMS is to assess conditions periodically to provide data to

- Update network improvement programs
- Assess the structural integrity and possible failure of facilities
- Schedule rehabilitation and maintenance as indicated by these evaluations and updated predictions
- Check and update performance predictions
- Improve prediction models
- Provide a basis for evaluating construction and maintenance techniques

6.1.2 In-service condition deterioration process

Traditionally, infrastructure design practices have considered initial condition, load, and material properties as the primary input variables for structural design without taking into account the effects of environment and material degradation over time. Such an approach is inadequate, and actual life is generally less than the predicted life. An early benefit of performance evaluation of highways was the identification of important environmental factors [Uddin 95a]. Figure 6.1 illustrates related factors in the case of highway and airport pavements. It is important to identify key factors, in addition to usage and aging, that affect the in-service condition deterioration to develop a meaningful program for in-service monitoring and evaluation.

Table 6.1 lists such factors in five major categories: load, environment, material degradation, construction quality, interaction effects, and other mechanisms. Table 6.1 also lists primary and secondary condition deterioration mechanisms associated with possible surface defects, deformation, cracking, and failures. Examples of catastrophic failures and their causes are also listed.

The primary factor in most structural deterioration and failure is load or load-accelerated distress initiated by environmental factors and material degradation. Therefore, it is important to record the loads acting on a facility, including the magnitude and repetitions. Examples of load considerations for a bridge evaluation are: (1) dead load (weight of the bridge structure without traffic); (2) primary live load (for example HS 20 Truck used by AASHTO and shown in Figure 6.2); and (3) sec-

Figure 6.1   Key factors influencing the performance of highway and airport pavements.

TABLE 6.1 Possible Condition Deterioration Mechanisms for Consideration in Infrastructure Evaluation

Deterioration manifestation	Deterioration Mechanisms					
	Load/ usage	Environment	Material degradation	Construction quality	Interaction	Other
1. Surface defects	Secondary	Primary	Primary	Secondary	Material/ environment and extended by load	Man-made defects, maintenance patches
2. Deformation	Primary	Secondary	Primary	Secondary	Material/ environment/load	Loss of foundation support, no maintenance
3. Cracking, disintegration, and breaks	Primary	Primary	Secondary	Secondary	Load/environment/ material degradation	Corrosion damage, accidents, no maintenance
4. (a) Failure aging/ inadequate structural capacity *or* retirement	The facility is structurally deficient because the limiting threshold values of (1) surface defects (2) deformation and (3) cracking and disintegration are exceeded. *Or* the facility is retired because it is functionally obsolescent.				Nature	Capacity and safety considerations, no maintenance
4. (b) Catastrophic failure	Primary causes: earthquakes, floods, freeze/snow, ice, tornadoes/cyclones, wind storms, foundation and soil failure, and sink holes				Poor construction quality/ design deficiency	Fire, arson, terrorist act, or accidents

Figure 6.2 The truck loading design for highway bridges. [AASHTO 92].

ondary live load (for example wind, thermal forces, braking load, earth pressure, buoyancy, centrifugal force in curves, stream flow, ice pressure, earthquake loads, and impact load).

Water and sewer pipe, generally laid under roads or buried underground within the road right-of-way, are subjected to the dead load of fill and pavement and the live load of passing traffic, earth pressure and hydrostatic water from all other sides, and the pressure and weight of water or waste/sewer water transported through the pipe. Just like a bridge or a building, the condition of these pipelines are dependent on the material of construction, the loads, and deterioration, such as corrosion. Large-diameter sewers and water tunnels made of bricks and concrete may experience structural failures as a result of erosion of bricks and mortar, and loss of support of the surrounding material. Internal visual and video-based inspections provide the best evaluation of overall integrity and location of problem areas.

As discussed in Chapter 5 and noted in Table 6.1, environmentally induced cracking can be a primary cause of failure in concrete and metal structures. This may then lead to premature failure, such as corrosion and fracture of steel components, loss of water by leakage through breaks in water lines, failure of a high-pressure gas pipeline, plane crashes, shut down of nuclear power plants, and contamination of soil and water sources. Freeze-thaw effects can significantly lower the load-carrying capacity of pavements.

Seasonal changes in moisture and temperature can induce tensile shrinkage stresses in the structure and reduce the subsoil load-carrying capacity. However, it should be recognized that interaction of a load with one or more environmental mechanisms is more critical for condition deterioration than environmental mechanisms alone. Heavy usage and load repetitions will in all cases accelerate the damage caused by environmental factors and material degradation. Thermal cracking in slabs and walls constructed with concrete or brick can be caused by tensile stress due to high temperature differential from one surface to the other [Uddin 83, Ho 95] and low-temperature shrinkage cracking of asphalt pavement [Haas 73, Harvey 94] can be caused by low temperatures. Potholes appear more often on asphalt roads during the spring thaw season, particularly in the presence of heavy traffic. This represents a good example of interaction of all three mechanisms: load repetitions, weakening of sublayers and roadbed soil, and thaw (environmental) conditions.

Primary and secondary mechanisms of condition deterioration dictate the plan for a cost-effective and efficient data-collection program for condition assessment and in-service evaluation. Table 6.1 provides guidelines for these mechanisms. These observations and mechanisms are well established for road and airport pavements constructed with asphalt, concrete, or without paved surface [FAA 82, SHRP 93, Eaton 87].

6.1.3 Approaches for in-service evaluation

In-service evaluation of a facility involves three steps:

- *Selecting sample sections* with inventory data and historical information on usage and load.
- *Monitoring condition* via measurements or observations recorded in a useful format.
- *Processing and evaluation* via manipulation and interpretation of the data collected to provide an overall condition rating and to make a judgment based on the results.

Evaluation involves the assessment of facility conditions with respect to the *user,* the *function,* and the *structural condition.*

6.1.3.1 User evaluation Common factors in user evaluation are convenience, comfort, health, safety, aesthetic aspects, and overall satisfaction with the quality of service. These are generally subjective in nature and often depend on visual observation or opinion. Examples include: (1) Present Serviceability Rating or PSR used for highways based on riding quality [HRB 62]; (2) sufficiency rating for bridges [Hudson 87]; (3) safety rating of air travel; and (4) citizen's complaints of poor water pressure and/or accumulation of garbage on streets. The subjective rating scale is generally numeric, ranging from zero (the worst or failed condition) to a specified maximum of 5, 10, or 100 for excellent condition.

6.1.3.2 Functional evaluation. Functional evaluation is the effectiveness of a facility in fulfilling its intended functions and is often similar to user evaluation. Major concerns are safety, size adequacy, sufficiency based on capacity and demand requirements, serviceability or quality of service, and physical appearance. Use of objective measurements are preferred because they lead to more productivity and better repeatability. Examples include: (1) Present Serviceability Index (PSI) based on pavement roughness measurements used for highways; (2) Pavement Condition Index (PCI) based on distress measurements used for highways and airports [FAA 82, Shahin 90]; (3) condition index for bridge decks based on detailed distress mapping and analysis; (4) track quality index for railroad tracks based on objective measurement of track geometry, cross slope, and alignment [Fazio 80]; (5) water quality and health hazard; (6) performance index for water lines and gas lines based on leakage analysis from supply and consumption records or nondestructive leakage testing of utility lines; (7) level-of-service and congestion for transportation facilities; and (8) hazards and safety problems due to reduced skid resistance, hydroplaning, and poor drainage on roads. Where possible, the condition or performance indices are based on objective measurements of selected performance indicators, or an analysis of maintenance records.

6.1.3.3 Structural evaluation.   Structural evaluation is used to assess structural integrity by conducting tests and structural analysis of test data. Structural evaluation includes load-carrying capacity, structural integrity, seismic or fire hazard, and safety of users. Example are: (1) Structural Capacity Index, used for highways based on pavement-deflection testing and load-carrying capacity analysis [Haas 94, Uddin 95b]; (2) structural rating and remaining life for bridge structures based on nondestructive deflection testing and structural fatigue analysis; (3) track modulus index for railroad tracks based on nondestructive deflection testing, or remaining life based on fatigue testing; and (4) frequency of breakdowns and failures of utility lines. Structural condition indices, based on objective measurements, indicate structural integrity and load-carrying capacity. Nondestructive methods are desirable in most cases.

Harty and Peterson [Urban 84a] point out that reliable information on the condition of infrastructure enables local officials to accomplish the following:

- To rate and rank facilities according to their current physical condition and performance.

- To determine deterioration rates and the best time to undertake maintenance for each facility. This will reduce the likelihood that maintenance will be undertaken too early or too late.

- To build a more effective constituency for capital upkeep by providing the public with reliable information on the current facility condition and projections of the consequences of underspending on capital facility preservation.

6.1.4 Data details and data collection frequency

The first basic requirement for condition data collection is a proper location reference system, as outlined in Chapter 5. It is desirable to have a common location-referencing scheme across an agency so that data of all types can be linked among inventory, construction, maintenance, and so forth. It is imperative that evaluation measurements be properly indexed by section, subsection, and date for efficient data management. The selection of data types and details can be linked to the information quality level (IQL) concept discussed in Chapter 5. The precision and extent of data items depend on the information quality level being sought.

Evaluation databases vary from a few mandatory elements to a comprehensive list of data elements, depending on the functional level of IMS (whether it involves network-level planning or detailed project-level design). As shown in Figure 2.4, the data details and system complexity increases as one moves from the network-level to project-level IMS functions. At the network level, high productivity and low cost are essential; however, the data must provide insight into M,R&R needs of the network. Evaluation at the project-level is more detailed, involving causes of deterioration and subsequently predicting appropriate M,R&R strategies. For research objectives, very detailed and more frequent measurements and detailed analyses are required.

The next step is to select the condition attributes that will be evaluated. The final selection of the factors should be made by a group of experts familiar with local infrastructure conditions. Carefully written definitions of the data items should be prepared and reinforced with photographs. If possible, the definitions should include instructions for rating the severity and measuring the extent of each condition attribute.

There are no absolute standards for the evaluation frequency. Guidelines can be based on available resources and data-processing speed, as recommended in Table 5.2 for different decision support system (DSS) components. Table 6.2 shows the recommended collection frequency for project-level usage and evaluation data. At the network level, major roads and bridges should be evaluated annually, but no less than once in three years. Priority should be given to aged facilities or those exhibiting problems. One example is the sewer system in St. Louis, Missouri, which is reportedly among the oldest in the U.S.,

TABLE 6.2   In-Service Evaluation Frequency and Database Update

DSS component	Database Update		Database file
	Mandatory	Desirable	
Usage (Current usage record)	Annually	Monthly and seasonal in addition to yearly total	Project
Condition evaluation	Sample once every 1–3 years	Annually (on 100% network)	Project

with some parts dating to 1850 [Collins 95]. Since most sewers are built under road pavement or in right-of-ways, sewer failure often results in cave-in of the overlying road with a huge repair cost. To comply with clean water regulations, a project was designed to implement an overflow regulation system and eliminate sewage overflows during high-river stages of the Mississippi River. As a part of a major structural rehabilitation project, criteria were established to identify those sewer sections most at risk, to define pumping station modification needs, and avoid external hydrostatic pressure [Collins 95].

Because condition assessments are costly, trouble-prone facilities should receive the most frequent assessment. Local agencies should identify the facilities where problems are most likely to occur, or where failures will have the most costly consequences. Other facilities can be inspected less often and/or on a sampling basis [Urban 84a, Urban 84b, Urban 84c].

6.1.5 User interface and system operation issues

The rate of physical deterioration is often a critical element of in-service evaluation, and it is important to keep records that are consistent from year to year. Database entries on usage should be updated regularly as new data become available, and in-service monitoring and evaluation database files should be maintained by survey date so that all historical evaluation data files are readily accessible for M,R&R needs assessment, improvement of prediction models, and other project-level applications.

6.2 In-Service Evaluation of Physical Assets

Typical examples of recommended evaluation data elements for various infrastructure groups are shown in Table 6.3. As constructed quality affects facility life, assessment immediately after construction should be included in the evaluation database.

6.3 Technologies for In-Service Monitoring and Evaluation

6.3.1 Overview

Monitoring data whether traditional, nondestructive, or innovative may be grouped into the following categories.

6.3.1.1 General automated records.

Some data are available in electronic and on-line databases as well as published documents for the location of interest. A good example is the weather and climate data recorded on airports and other local weather channels available from

TABLE 6.3   Examples of Infrastructure In-Service Monitoring and Evaluation

| | In-Service Condition Monitoring and Evaluation Data Elements | | |
Facility	User evaluation	Functional evaluation	Structural evaluation
(1) Highways, roads, streets, parking areas	Present Serviceability Rating (PSR), ride quality, vehicle operating costs, user satisfaction ratings based on congestion and pollution, traffic data	Present Serviceability Index (PSI), International Roughness Index (IRI), Pavement Condition Index (PCI) based on distress data, Pavement Condition Rating (PCR)	Deflection testing, remaining life, structural capacity index
(2) Airports: pavements, buildings, aviation facilities	Ride quality, user satisfaction ratings for other facilities, aircraft operations data	PCI (pavements), Ride Quality Index (RQI); see (8) for buildings, air-traffic control facilities, airport lighting, etc.	Deflection testing, remaining life, structural capacity index; see (8) for buildings and other facilities
(3) Bridges: interchange, overpass, underpass flyover, rail/river crossing	Ride quality, user satisfaction ratings, traffic data	Bridge inventory and appraisal ratings, sufficiency rating based on inventory and appraisal ratings, deck rating	Vibration and seismic testing, nondestructive evaluation, load rating, remaining life
(4) Railroad: terminals, track, bridge	Service quality and efficiency; user satisfaction ratings, train operations data	Track Quality Index (TQI); see (3) for bridges; see (8) for buildings and other facilities	Track modulus, remaining life; see (3) for bridges; see (8) for buildings and other facilities
(5) Water: treatment plants, storage, supply	Health hazard, water treatment, user satisfaction, water pressure, and consumption	Number of breaks per year, leakage per year, number of repairs	Failures, damage to roads and other structures, remaining life
(6) Waste water and sewer	Health hazard, water treatment, user satisfaction, discharge data	Number of breaks per year, leakage per year, number of repairs	Failures, damage to roads and other structures, remaining life
(7) Solid-waste facilities	Health hazard, fire hazard, odor and treatment, user satisfaction, waste load	Spilled trash, missed collections, soil and water contaminations	Failures, damaged and inadequate collection and disposal facilities
(8) Buildings: multipurpose complexes	Complaints, discomfort, user satisfaction ratings, live load and occupancy data	Interruptions in utility services, heating and cooling effectiveness, air quality, condition ratings	Structural and ground failures, posting, remaining life, safety hazards
(9) Power-supply facilities	Fire hazard, interruptions in service, user satisfaction, power generated and distributed	Frequency of service breaks, power surges, condition ratings of poles and cables; see (1) for roads; see (8) for buildings and other facilities	Failures, remaining life; see (1) for roads; see (8) for buildings and other facilities

NOAA [NOAA 94] and the Internet for most of the United States. An airport will have aircraft traffic data available through records of air-traffic control. Traffic history, loading, axle configuration, and volume data for roads and bridges are sometimes collected and stored in central transportation agency databases.

6.3.1.2 Visual inspection. Visual observation and inspection is the most widely used method of condition monitoring. Manual visual inspections are labor-intensive, expensive, and subject to inspector's judgments. However, in some cases they are necessary, such as postearthquake evaluation [ATC 93] or after other catastrophes (hurricane, floods). Costs can be reduced through sampling and/or summary methods, such as windshield surveys.

6.3.1.3 Photographic and optical methods. Photographic and optical methods include videotape recording, 35-mm photologging, borescope inspections, internal pipe video, and other optical methods with permanent records [Hudson 87a, Hudson 87b]. The records may be obtained from selected positions, slow walk speed, or high-speed dedicated vehicles. Photo records must be read or interpreted to produce data. Automated pattern recognition techniques that can identify and quantify data are still in the development stage. Video-based methods are being used successfully for traffic studies.

6.3.1.4 Automated geometrical data collection. Automated measurements of geometrical characteristics can be cost-effective for network-level evaluation, for example, longitudinal and cross-fall deficiencies on paved surfaces and slabs, and track geometry [Hudson 87b, Fazio 80]. Such data collection can be especially useful and efficient on roads, railroad tracks, and pipelines.

6.3.1.5 Nondestructive structural evaluation methods. Structural integrity of a facility can be checked using nondestructive evaluation (NDE) methods, which do not subject the facility to actual loading or destructive testing. These NDE methods include seismic evaluation, such as wave propagation; vibration methods, such as modal analysis; acoustic and ultrasonic methods; dynamic deflection testing; x-ray diffraction techniques; electromagnetic methods, and electrical resistivity methods [Hudson 87a, Hudson 87c, Hudson 87d, Matzkanin 84, Metzger 93, Uddin 94, Uddin 96a, Uddin 96b, NDT 96]. Corrosion damage in steel and concrete can be estimated by electrical resistivity methods [Chaker 96, Bridge 96a]. Other noncontact and nondestructive testing (NDT) technologies for structural evaluation include radar, infrared thermography, laser, high-speed video and related optical methods, Moire technique, and ultrasound sensors [Uddin 94, NDT 96].

6.3.1.6 "Smart" sensors. "Smart" sensor technologies rely on the embedment of fiber optics, ultrasound sensors, or piezoelectric sensors for life-long monitoring and evaluation of structures [Ansari 96, Metzger 93]. Fiber optic technology offers a potential application to monitor deformation and stress responses to loads.

6.3.2 Destructive testing and evaluation

Although the in-service condition can be evaluated by surface or NDT observations, it occasionally becomes necessary to remove portions of a facility to ascertain more specifically where problems are occurring and why. This is called *destructive testing* because a portion of the original structure is removed and must be repaired or replaced. In general, such procedures are used where there is evidence of unexpected deterioration; however, destructive methods have been used on full-scale tests of roads, dams, pipelines, underground tanks, and pile foundations as well. For example, they have been used on the AASHO Road Test [HRB 62] and form an important part of the SHRP-LTPP study [SHRP 89a, SHRP 89b]. Surface defects can be used as general guides to the underlying conditions; however, it is often necessary to open a facility internally to determine the true position and cause of failure for a completely reliable analysis.

Minimum destructive testing techniques include coring and coupon sampling of structural members, such as bridge decks. On highways and airports or industrial floors, destructive testing is conducted by coring or cutting into the facility and removing samples of the various layers, examining the samples in the field, and then testing them in the laboratory. In some cases, full destructive tests are carried out until the structure yields or fails; for example, a pile load test. Such tests are performed on a sample basis because of the time and cost involved in the destructive testing. The results then are inferred to the remaining units.

6.3.3 Traffic, usage, and/or occupancy

An important part of in-service data is measurement of traffic or other usage factors, such as volume of flow or number of take-offs and landings. Generally, each type of infrastructure has its own type of usage. For airfields, this includes the number of take-offs and landings and the number of individual enplaned counts. These data are recorded by commercial airlines and by the Federal Aviation Administration (FAA) for airplane traffic.

For highways, the data is collected on a sampling basis using traffic counters at specific locations for several days per year. These data are then expanded to estimate the annual average daily traffic for each highway location. Load information is estimated by taking a sample weighing of axle loads by various classes of trucks and by doing automatic classifications of vehicles in various classes, passenger cars, light trucks, three-axle trucks, and so forth, through five-axle tandem trucks. There is a wide variety of automatic data-collection devices used for these data.

Flow of sewage and water are measured by flow meters in selected pipes and then expanded to the entire flow-pipe system. Flow capacity and quantity are also recorded at water plants and sewage-treatment plants. Sampling is used effectively.

For parks and other such facilities, entrance access is usually controlled and recorded. These data can be used for IMS purposes.

6.3.4   Evaluation methods and applications

Table 6.4 lists several evaluation methods, with examples of application, to evaluate various types of infrastructure.

6.4   Inspection, Photographic, and Optical Evaluation

Assessment of physical condition and defects is an integral part of effective in-service evaluation. Video and 35-mm continuous photography methods are being used for pavement inventory and distress-data acquisition in pavement-management applications. Photologging techniques have been used by several agencies for geometric inventory purposes and safety evaluation. Videotape recording is also very useful for traffic data collection and inspecting bridges, tunnels, water mains, sewers, and other structures that are hard to access by a human body. The use of photographic methods and video provide a permanent record of the image at the time of evaluation, and comparison of these records over time can be useful in interpreting condition deterioration rates and mechanisms. In some cases, accuracy and information on some observed distress are lost when reducing the condition-deterioration data from photo records, because the interpretation is based on two-dimensional images and current pixel size.

Continuous 35-mm strip photographic techniques for distress survey of highway and airport pavements and right-of-way inspection have been used in the United States since 1986 and in France and Japan for several years [Hudson 87b]. Video technology for recording

TABLE 6.4   Application of Traditional and Innovative Technologies for In-Service Evaluation and Condition Assessment of Infrastructure Facilities

Technology	Applications	Infrastructure
Visual inspection	Manual record of observed distresses and defects	All types of infrastructure assets
35-mm continuous photography	Permanent record of physical condition, inspection, distress, and defect data acquisition	Road and airport pavements, all other assets
Video recording	Permanent record of physical condition, inspection, distress, and defect data; road vehicular traffic data	Road and airports, railroad tracks, sewer and water lines, bridges, all other assets
Seismic, deflection, and vibration test methods	In situ material characterization, structural integrity evaluation, seismic risk assessment	Road and airport pavements, bridges, all other structures
Radar	Structural integrity evaluation, delamination, layer delineation, voids and moisture damage, dielectric constant properties	Road and airport pavements, bridges, all other concrete and masonry structures
Infrared thermography	Temperature measurement, detection of moisture leakage, delamination, defective areas, stress mapping	Road and airport pavements, bridges, all underground assets, leakage and breaks of water and sewer lines
Acoustic and ultrasonic testing	Cracking and defect detection, longitudinal roughness and rut depth measurement on paved surfaces, stress, strain, displacement measurement	Road and airport pavements, bridges, underground utilities, all other structures
Laser	Cracking and defect detection, profile measurement, longitudinal roughness and rut depth measurement, faulting of transverse joints, surface texture and displacement measurement	Road and airport pavements, bridges, other structures
Fiber optics	Stress and strain measurement	Bridges, tunnels, "smart" structures
Piezoelectric sensor	Stress, strain, traffic measurement	Road and bridges
Electrical resistivity	Detection of surface and subsurface material changes, corrosion damage	Soils characterization, concrete corrosion
Electromagnetic conductivity	Detection of cracking, voids, and subsurface defects	Bridges, tunnels, other structures
Topographic surveying, GPS, gyroscope	Location identification, alignment, longitudinal and cross falls, routing, real-time tracking	Road and airport pavements, bridge decks, railroad tracks, utilities, water and sewer

images of the pavement surface and collecting distress data (including interpretation) is now available commercially [Hudson 87a]. Photographic methods are becoming more popular because of the technological advances in computer-based digital image processing. For example, videotaped images of 33,600 km (21,000 mil) of main railroad track and a GIS database in the Engineering Facilities Management System (EFMS) were used by the Union Pacific Railroads to assess the performance of rail track, to program on-time maintenance, to improve safety of operations, and to assist in real estate management [Gerard 91]. Video imaging is also being used for roads, airport pavements, and bridges where manual on-site inspection is difficult and risky in highly trafficked sections. Video also provides remote assessment of physical condition, structural integrity, and emergency maintenance needs for buildings and sewer and other pipeline systems.

6.5   Nondestructive and Noncontact Structural Evaluation

Nondestructive and noncontact testing and evaluation techniques are the primary methods used for structural integrity and load-carrying capacity evaluations. These techniques are listed in Table 6.4. In this section, some of the successful and promising methods are discussed in more detail.

6.5.1   Vibration, seismic, and ultrasound methods

Several types of vibration and seismic equipment developed for industrial vibration problems and earthquake engineering are now used for structural evaluation of pavements, foundations and structures. Stress wave propagation using spectral-analysis-of-surface waves (SASW) [Heisey 82, Nazarian 83, Aouad 96], CEBTP's transient dynamic response technique [Uddin 94, Hertlein 96], and acceleration monitoring for structure modal analysis [O'Leary 96] have been applied with success. The advantages of such techniques are the elimination of holes in the structure, tests results under realistic field conditions without altering the mechanical state of stress, and reduced traffic disruption.

6.5.1.1   Principles of wave-propagation tests.   The wave-propagation test methods primarily measure wave velocity at or near the surface of a structure subjected to a small transient or vibratory force. Generally, accelerometers and/or velocity transducers (geophones) are

used to measure the surface response. The tests are used to measure the velocity of compression waves (P-wave), shear waves (S-wave), or Rayleigh waves (surface wave or R-wave), depending upon the mode of excitation forces [Lysmer 70, Abbiss 81]. Figure 6.3 illustrates these wave forms. The wave velocities are related to material moduli. These low strain-amplitude nondestructive tests can provide an estimate of linear elastic material properties. The subsoil properties can be used to evaluate liquefaction potential and soil support.

Steady-state vibration techniques have been traditionally used for stress wave testing using time-domain analysis. Shear wave tests are

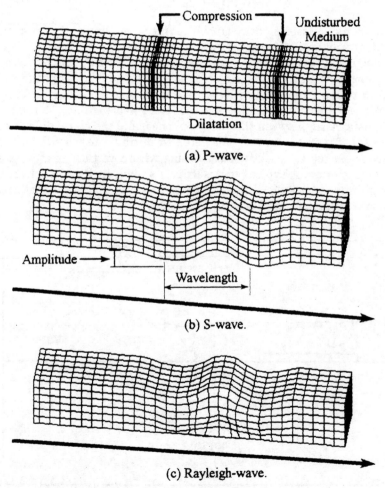

(a) P-wave.

(b) S-wave.

(c) Rayleigh-wave.

Figure 6.3 Illustration of stress wave propagation.

particularly useful for soils and foundations because the shear wave velocity (V_S) is directly related to the shear modulus (G) by:

$$G = \rho(V_s)^2 \qquad (6.1)$$

where ρ = mass density, kg/m^3 (unit weight/acceleration due to gravity).

Young's modulus (E) for a homogeneous linear elastic material can be calculated from G by:

$$E = 2G(1 + \mu) \qquad (6.2)$$

In the case of the p-wave, the modulus depends on the Poisson's ratio, μ:

$$G = \rho(V_p)^2[(1 - 2\mu)/2(1 - \mu)] \qquad (6.3)$$

The dynamic Poisson's ratio can be found from the ratio of p-wave and shear wave velocities using Eqs. (6.1) and (6.3).

Rayleigh waves travel on the surface with an elliptical particle motion at a velocity (V_R), close to that of a shear wave. The amplitude of Rayleigh waves attenuates with depth, and most of the energy is found within one wavelength from the surface. The wavelength, L_R, is related to the distance, d, between the two geophones, and the phase difference (degrees) between the two signals received during a test in which the surface is excited by a vibrator oscillating with a vertical steady-state harmonic motion. A typical test setup is shown in Figure 6.4. The Rayleigh wave velocity is assumed to correspond to that of the material at a depth of one-half wavelength.

Figure 6.4   Rayleigh wave test. [after Abbiss 81].

Figure 6.5   Schematic of the SASW test setup. [Nazarian 83].

6.5.1.2 SASW method. The SASW method uses frequency-domain analysis of Rayleigh wave propagation to determine pavement moduli and layer thicknesses [Nazarian 83]. Recently, it has been used to estimate the depth of bridge foundations [Aouad 96, TRB 95]. The test setup is shown in Figure 6.5. The excitation source is a hammer blow that applies a transient vertical impact, which generates a group of Rayleigh waves of various frequencies. The propagation is monitored by two geophones on the surface. The output of the receivers are simultaneously recorded on a Fourier spectral analyzer that decomposes the transient wave forms into a group of simple harmonic wave forms that are analyzed individually to determine R-wave velocity and wavelength associated with each frequency.

6.5.1.3 Transient dynamic response method. The transient dynamic response method has been used to locate voids under concrete pavements [Uddin 94] and to study the integrity of shaft foundations [Hertlein 96]. A hammer equipped with a load cell is used to generate small stress waves and the response is measured by a geophone. The time-domain data is converted into the frequency domain, which is interpreted to provide information on dynamic stiffness and concrete quality.

6.5.1.4 Impact echo (IE) and ultrasonic pulse velocity (UPV) methods. The Impact Echo (IE) method is a popular and common ultrasonic technique in which a hammer is used to excite the surface axially and the response of the reflected waves is measured by an accelerometer.

Reflection   Reflection
from slab/void   from
interface   bottom of slab

* Reflection from bottom occurs at a
lower frequency than that from the
shallower slab/void interface

POINT-BY-POINT   **SCANNING IE TESTS ON TOP**
IMPACT ECHO (IE) TESTS   **OF A BRIDGE DECK**

Figure 6.6   A schematic of Impact Echo test setup. [after Olson 96].

The low-strain compression waves travel through the thickness or
depth of the structure and are partially or completely reflected by
changes in the conditions, such as varying cross sections, concrete
defects, voids, delaminations, and so forth. This test is used for concrete structures, including beams, columns, foundations, pavements,
and bridge decks, to evaluate structural integrity, defects (such as
voids, honeycombing, cracks), and delaminations [Olson 96, Hertlein
96]. Recently, advanced scanning techniques have been developed in
Germany [Schickert 96].

Figure 6.6 shows a schematic of the IE method and a mobile IE
scanner developed for bridge decks [after Olson 96]. The Ultrasonic
Pulse Velocity (UPV) method is based on compression wave analysis if
two surfaces are accessible, as shown in Figure 6.7 [after Olson 96].
Currently, a superscanner is being developed by U.S. Army Corps of
Engineers [Alexander 96] to detect delamination of bridge decks.

Parallel seismic test.   The parallel seismic test is an effective method
to evaluate piles, drilled shafts, and deep foundations [Hertlein 96].
The test is very suitable where the head of pile or shaft is inaccessi-
ble. It has been successfully used to measure the depth of bridge foundations [Aouad 96, TRB 95]. Figure 6.8 illustrates the test setup
[Hertlein 96]. A bore hole is made in the ground near to, and parallel
with, the foundation. The bore hole is lined with a PVC pipe and filled
with water to provide an acoustic response that will be monitored by
hydrophones.

UPV SCANNING FOR VELOCITY/TOMOGRAPHY TESTS

Figure 6.7 Ultrasonic Pulse Velocity (UPV) test configuration. [after Olson 96].

Figure 6.8 The test configuration for parallel seismic test. [after Hertlein 96].

6.5.2 Dynamic deflection equipment

Commercially available dynamic deflection equipment measures and records the dynamic response of a road or airport pavement subjected to steady-state vibratory or transient loading forces. Examples are Dynaflect, Road Rater, and falling weight deflectometer (FWD). Improvements have been made to automate the testing sequence and provide on-board data processing. Table 6.5 shows guidelines for the sampling rate.

TABLE 6.5   Recommended Sampling Rates and Methods for Pavement Nondestructive Evaluation [Paterson 90]

Application	Sampling rate	Information quality level (IQL)	Evaluation method
Network planning	1–5% network length (Stratified random sample)	IV or III	Dynamic deflection Construction record DCP profiles Visual survey
Network programming	0.3–1.0 km intervals (min. 5 points per section) JRP: At slab center and joints	III or II	Dynamic deflection Deflectograph deflections
Project design	20–200 m (min. 5 points per section) each wheelpath. RP: 20–40 m in outer wheelpath	Minor roads: III or II Major roads: II or I	Dynamic deflection Deflectograph deflection DCP profiles + SN calculation [+field sampling + material testing as required]
Research and special investigation	3–20 m intervals RP: every joint or crack	II or I	NDT deflection + field sampling + material testing

Notes: Dynamic deflection devices = Dynaflect and FWD (falling weight deflectometer); NDT = nondestructive testing; DCP = dynamic cone penetrometer; RP = rigid pavements; JRP = jointed rigid pavements; SN = structural number

6.5.3   Time-domain reflectometry (TDR)

The time-domain reflectometry method was originally developed by the power and telecommunications industries to locate breaks in cables, and the method has been extended to measure material dielectric constant, concrete material properties, discontinuities in concrete and rock [Su 96], and to monitor unstable slopes [Kane 96].

6.5.4   Noncontact methods

6.5.4.1   Ground-penetrating radar technology.   Ground-penetrating radar used for civil engineering applications is an adaption of the vehicle-mounted mine detection radar developed for the United States Army. This technology has been used to identify the subsurface structure of soil, and to locate underground utilities, pipelines, and tunnels. With improved antenna and other hardware developments, the technology has been applied to pavement. Radar testing devices are available commercially for identifying and locating voids under concrete pavement, locating steel reinforcement, evaluating effectiveness of grouting work under concrete pavements, and other related applications [Uddin 94, Uddin 96b]. The interpretation of the response outputs is the most critical element in correct and effective use of radar. The test monitors the dielectric constant profile, which is then used to estimate layer thickness and material properties. The advent of color output monitors has improved the value of radar tests, but the true effectiveness of these methods remains to be proven.

A radar unit emits a short pulse of electromagnetic energy and examines the reflected energy from that pulse. In the case of a pulse fired into a pavement structure, the electromagnetic wave travels until it meets a discontinuity caused by a change in the dielectric constant of the material in the path of the wave. A portion of the wave is reflected by this discontinuity and a portion continues to travel through subsequent layers.

The amount of energy reflected at the discontinuity is a function of the wave impedance of the two materials. At the interface between materials with similar dielectric properties, such as two lifts of an asphalt concrete pavement, most of the energy passes through the interface and very little is reflected back to the transmitter. Conversely, where the difference in dielectrics is significant, such as in an asphalt layer over concrete or over granular base, much of the energy is reflected to the transmitter and very little is passed to the next layer. This reflection phenomenon is the theoretical basis for the various radar signatures produced by different subsurface anomalies [Pulse 94].

Data can be collected at up to normal highway speed at a rate of 50 data points per second. The time from the firing of the pulse to its return to the antenna is very accurately timed. This pulse travel time is a function of the thickness of each layer and a material property known as the *dielectric constant*. The dielectric constant is a measure of the ability of the material to pass electromagnetic radiation through it. Thickness of each layer can be estimated by the expression:

$$D_n = 6t_n/[E_n]^{0.5} \qquad (6.4)$$

where D_n = thickness of layer n,

t_n = the two-way travel time (in nanoseconds) of the pulse in layer n, and

E_n = the dielectric constant of layer n.

The methodology is described in *ASTM D4748-87, Standard Test Method for Determining the Thickness of Bound Layers Using Short Pulse Radar* [ASTM 91]. The data is acquired using a short-pulse radar with a center frequency of 1 GHz. Longitudinal position is determined by a transmission-mounted encoder. The data is digitally stored in a 386 or higher microcomputer for processing, and is displayed in color on a monitor for an expert analyst to review [Pulse 94].

6.5.4.2 Acoustic and ultrasonic testing.

Nondestructive sonic testing has been applied to detect delamination, honeycombing, and other discontinuities in concrete slabs. In the pulse-echo method, a compression wave is sent through the material and the operator monitors the reflection pattern of the wave caused by discontinuities in the material. Acoustic methods have been particularly useful to evaluate cracks, corrosion, and other defects in concrete [NDT 96].

In recent years, noncontact ultrasonic sensors have been used for pavement evaluation for longitudinal profile and rut-depth measurements [Haas 94]. Ultrasonic sensors are, however, sensitive to environmental factors like humidity and moisture and may produce bad data under some conditions [Hudson 87a].

6.5.4.3 Laser testing.

Laser technology holds promise for several commercial applications in pavement evaluation. It has been commercially applied for counting transverse cracks and measuring joint faulting, longitudinal roughness, and surface texture [Hudson 87a, Hudson 87b]. Laser-based pavement evaluation equipment should be used more often in the future because of its high level of automation, leading to greater productivity and reduced operating constraints (like traffic interference, environmental factors, and speed dependency).

6.5.4.4 Thermal infrared photography.

Thermal infrared photography can be used to locate defects and voids in concrete structures, bridge decks, and buried pipes [Uddin 96b, Delahaza 96, Del Grande 96]. Infrared thermography is a noncontact, noninvasive, means of obtaining visible images from the invisible heat energy emitted by an object. Television-like images are produced with gray-scale variations, or different colors, representing various temperatures. Thermography has limitations, as do all temperature-sensing and measurement techniques. The observed radiometric temperature is affected by such things as the object's absolute temperature, the ambient temperature, the object's emissivity, the radiant energy of nearby surroundings, the atmospheric filtering of energy, and the distance from the object to the scanner, all of which require that the thermographic information be adjusted [Entech 94].

An infrared imaging system consists of an infrared scanner, which records primarily infrared radiation, and real-time computer display monitor (similar to a portable television with a small screen coupled to a microprocessor). The scanner unit converts the radiated heat that passes through the optics into an electronic signal. The signal produces a real-time thermal image on the computer screen composed of a gray scale with continuous tones ranging from black through gray to white. Areas of higher relative temperature appear lighter and areas of lower relative temperature appear darker. A color monitor and an additional microprocessor can also be used to display the thermal image. In this case, each color represents a particular temperature range.

The infrared scanning systems used in a concrete pavement study [Uddin 96b] allowed for rapid data collection and analysis. With this equipment it is possible to observe, quantify, and record the thermal image of the surface of an object whose temperature is between −20°C and 2000°C. The sensitivity of the imaging equipment is such that it has the capacity to detect surface temperature differences between two given objects to an accuracy of ±0.05°C at an ambient temperature of 30°C [Entech 94].

6.6 Combined Evaluation Data

In many types of infrastructure, more than one method of evaluation is useful. In the case of pavements, for example, roughness, cracking, and deflection may all be used. In such cases, the various response data may be combined or aggregated into one composite measure of overall response. A combined index is one useful method to accomplish this task.

6.6.1 Reasons for use of a composite condition index

A composite condition index (CI) is sometimes used to combine two or more evaluation attributes and to communicate summary evaluation information to administrators, elected officials, and the public. Figure 6.9 illustrates the levels of data that are appropriate to various levels of decisions for infrastructure management.

At the first, or disaggregated, level, involving specific activities and specific technical engineering decisions, detailed monitoring and evaluation data attributes of each section are required. The second level involves aggregation and decisions for network-level management. At this level, composite measures and combined indices are useful for establishing priorities for the selection of projects and M,R&R strategies. Composite measures are absolutely required at the third level, which involves administrative and political decisions. Personnel at this level are faced with large amounts of information, and therefore need aggregated data to portray the overall quality of the network and to project future quality as related to budgets. Generally, the CI

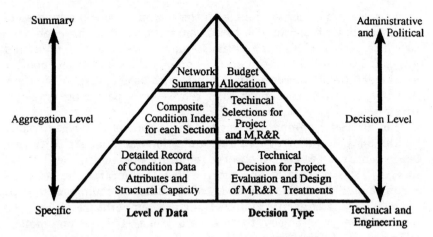

Figure 6.9   Data aggregation and level of decision. [after Haas 94].

is reported on a scale of 100 (excellent condition or new construction) to 0 (worst condition). The CI concept is also useful for developing performance equations that model the trend of condition deterioration and effect of M,R&R intervention, as discussed in Chapter 8.

6.6.2 Methods of developing CI

Each infrastructure facility requires a condition index that relates to the data being collected, the local conditions, and the objectives of the agency. There is no single best engineering or analytical formula for establishing a CI. Techniques for developing simple linear and nonlinear equations have been borrowed from the social-sciences field, using subjective data. A composite condition index combines two or more condition attributes and represents an aggregation of the different measures of condition; for example:

$$CI = W_1 C_1 + W_2 C_2 + W_3 C_3 \tag{6.5}$$

where CI = composite condition index;
W_1, W_2, W_3 = weighting factors for condition measures 1, 2, and 3; and
C_1, C_2, C_3 = values for condition measures 1, 2, and 3.

Other methods use the deduct-value concept. In this method a facility in excellent condition, such as new construction, is given a score of 100. Any defects or distress present results in some deduction, depending upon the defect severity and extent, which are subtracted from 100 to produce the CI at that time. Adjustments are made for multiple defects so the total deduct values can never exceed 100. This approach was used to establish the Pavement Condition Index (PCI) for pavements in the Washington Department of Transportation and was modified by the U.S. Army Corps of Engineers [Haas 94, FAA 82, Shahin 90]. This approach also was used to develop a sufficiency index for bridges [Hudson 87].

The Delphi technique can also be used to establish a composite CI by capturing expert opinion. It involves selecting a panel of experts who are asked to rank a series of hypothetical situations that describe a facility with different combinations of physical deterioration manifestations and structural capacity. The initial rankings are analyzed, and the results are presented to the panel for their assessment. An eventual convergence toward a common ranking is reached, and the information can then be statistically modeled to produce a composite CI [Haas 94].

6.6.3 Precautions in the use of composite CI

The key to developing a combined or composite index of infrastructure condition and service quality is to recognize the subjective nature of the problem and the associated techniques for quantifying subjective information. The methodology may be transferable, but not the specific models developed. Thus, the index will require calibration or redevelopment for each agency or geographic region.

Due to the aggregate nature of a composite CI, caution must be exercised in interpreting the index in selecting an M,R&R treatment. For example, the CI may produce a specific estimate at the network level for the selection of a treatment strategy for the M,R&R budget. But at the project level, using precisely detailed evaluation data, a particular defect may ultimately require the use of a different M,R&R treatment. This does not imply that composite indices are inapplicable at the project level, but it does mean that their primary use is to convey summary information at the network level [Haas 94].

6.7 Institutional Issues

Equipment, staff, garage, and office space are needed to develop and implement in-service monitoring, evaluation plans, and IMS evaluation database. The agency resource requirements and the system complexity for data collection and database development will be influenced by the size of the infrastructure network and the short-term and long-term goals for the data and M,R&R analysis. The amount of detail required for these various applications increases progressively from the overall summary statistics, through planning and programming, to design and research. The collection and processing of data will also be increasingly more complex and costly. In the face of limited budgets, compromises and wise decisions must be made.

References

[AASHTO 92] "Standard Specifications for Highway Bridges," 15th ed., American Association of State Highway and Transportation Officials (AASHTO), Washington, D.C., 1992.

[Abbiss 81] C. P. Abbiss, "Shear Wave Measurement of the Elasticity of the Ground," *Geotechnique,* vol. 31, No. 1, 1981, pp. 91–104.

[Alexandar 96] A. M. Alexander, R. W. Haskins, and D. E. Wilson, "Development of the Superscanner for the Detection and Mapping of Delaminations in Concrete Bridge Decks," *Proceedings, 2nd International Conference on Nondestructive Testing of Concrete in the Infrastructure,* Nashville, Tenn., 1996, pp. 39–47.

[Ansari 96] F. Ansari, Y. Libo, I. Lee, and H. Ding, "A Fiber Optic Embedded Crack Opening Displacement Sensor for Cementitious Composites," *Proceedings, 2nd International Conference on Nondestructive Testing of Concrete in the Infrastructure,* Nashville, Tenn., 1996, pp. 268–277.

[Aouad 96] M. F. Aouad, L. D. Olson, and F. Jalinoos, "Determination of Unknown Depth of Bridge Abutments Using the Spectral Analysis of Surface Waves (SASW) and Parallel Seismic (PS) Test Methods," *Proceedings, 2nd International Conference on Nondestructive Testing of Concrete in the Infrastructure,* Nashville, Tenn., 1996, pp. 147–153.

[ASTM 91] ASTM D4748-87, "Standard Test Method for Determining the Thickness of Bound Pavement Layers Using Short-Pulse Radar," *1991 Annual Book of ASTM Standards,* Vol. 04.03, pp. 557–562.

[ATC 93] Applied Technology Council, "Postearthquake Safety Evaluation of Buildings Training Manual," *ATC-20-T,* funded by the Federal Emergency Management Agency, Washington, D.C., 1993.

[Bridge 96a] "Calculating Corrosion," *BRIDGE Design and Engineering,* London, No. 3, May 1996, p. 78.

[Chaker 96] V. Chaker, "Measuring Soil Resistivity," *ASTM Standardization News,* vol. 24, No. 4, April 1996, pp. 30–33.

[Collins 95] M. A. Collins and C. T. Stude, "Masonry Sewer Rehab," *Civil Engineering,* vol. 65, No. 9, Sept. 1995, pp. 65–69.

[Delahaza 96] A. O. Delahaza, "Nondestructive Testing of the Concrete Roof Shell at the Kingdome in Seattle, Washington," *Proceedings, 2nd International Conference on Nondestructive Testing of Concrete in the Infrastructure,* Nashville, Tenn., 1996, pp. 256–267.

[Del Grande 96] N. K. Del Grande, P. F. Durbin, C. M. Logan, D. E. Perkins, P. C. Schaich, "Determination of Dual-Band Infrared Thermal Imaging at Grass Valley Creek Bridge," *Nondestructive Evaluation of Bridges and Highways, Proceedings SPIE 2946,* The International Society for Optical Engineering, Scottsdale, Ariz., pp. 166–177.

[Eaton 87] R. A. Eaton, S. Gerard, and D. W. Cate, "Rating Unsurfaced Roads," *Special Report 87-15,* U.S. Army Corps of Engineers, Cold Regions Research and Engineering Laboratory, 1987.

[Entech 94] EnTech Inc., *Infrared Thermographic Investigation for the Location of Highway Pavement Subsurface Anomalies,* report prepared for The University of Mississippi/ Mississippi Department of Transportation, May 1994.

[FAA 82] "Guidelines and Procedures for Maintenance of Airport Pavements," *Advisory Circular AC: 150/5380-6,* Federal Aviation Administration, Washington, D.C., 1982.

[Fazio 80] A. E. Fazio and R. Prybella, "Development of an Analytical Approach to Track Maintenance Planning," in *Transportation Research Record 744,* Transportation Research Board, National Research Council, Washington, D.C., 1980, pp. 46–52.

[Gerard 91]   S. V. Gerard, "UP Links Video and Graphic Images for Greater Clarity," *Railway Track and Structures,* April 1991.

[Haas 73]   R. C. G. Haas, "A Method for Designing Asphalt Pavements to Minimize Low-Temperature Shrinkage Cracking," *Research Report 73-1,* Asphalt Institute, 1973.

[Haas 94]   R. Haas, W. R. Hudson, and J. P. Zaniewski, *Modern Pavement Management,* Krieger Publishing Company, Malabar, Fla., 1994.

[Harvey 94]   J. Harvey, T. Lee, J. Sousa, J. Park, and C. L. Monismith, "Evaluation of Fatigue and Permanent Deformation Properties of Several Asphalt Aggregate Field Mixes Using Strategic Highway Research Program A-300A Equipment," in *Transportation Research Record 1454,* Transportation Research Board, National Research Council, Washington, D.C., 1994, pp. 123–133.

[Heisey 82]   S. Heisey, K. H. Stokoe II, W. R. Hudson, and A. H. Meyer, "Determination of In Situ Shear Wave Velocities from Spectral Analysis of Surface Waves," *CTR Research Report 256-2,* Center for Transportation Research, The University of Texas at Austin, 1983.

[Hertlein 96]   B. Hertlein and C. N. Baker, "Practical Experience with Nondestructive Testing of Deep Foundation," *ADSC,* The International Association of Foundation Drilling, March/April 1996, pp. 19–26.

[Ho 95]   D. Ho, "Temperature Distribution in Walls and Roofs," *Journal of Architectural Engineering,* vol. 1, No. 3, Sept. 1995, pp. 121–132.

[HRB 62]   "The AASHO Road Test, Report 5—Pavement Research," *Special Report 61-E,* Highway Research Board, National Research Council, Washington, D.C., 1962.

[Hudson 87]   S. W. Hudson, R. F. Carmichael III, L. O. Moser, W. R. Hudson, and W. J. Wilkes, "Bridge Management Systems," *NCHRP Report 300,* Transportation Research Board, National Research Council, Washington, D.C., 1987.

[Hudson 87a]   W. R. Hudson and W. Uddin, "Future Evaluation Technologies: Prospective and Opportunities," *Proceedings, 2nd North American Pavement Management Conference,* Toronto, Ont., 1987.

[Hudson 87b]   W. R. Hudson, G. E. Elkins, W. Uddin, and K. Reilley, "Improved Methods and Equipment to Conduct Pavement Distress Surveys," *Report No. FHWA-TS-87-213,* ARE Inc. Report for the Federal Highway Administration, April 1987.

[Hudson 87c]   W. R. Hudson, G. E. Elkins, W. Uddin, and K. Reilley, "Evaluation of Deflection Measuring Equipment," *Report No. FHWA-TS-87-208,* ARE Inc. Report for the Federal Highway Administration, March 1987.

[Hudson 87d]   W. R. Hudson, W. Uddin, and G. E. Elkins, "Smoothness Acceptance Testing and Specifications for Flexible Pavements," *Proceedings, 2nd International Conference on Pavement Management,* Toronto, Ont., 1987.

[Hudson 94]   W. R. Hudson and S. W. Hudson, "Pavement Management Systems Lead the Way for Infrastructure Management Systems," *Proceedings, 3rd International Conference on Managing Pavements,* National Research Council, vol. 2, May 1994, pp. 99–112.

[Kane 96]   W. F. Kane and T. J. Beck, "Rapid Slope Monitoring," *Civil Engineering,* vol. 66, No. 6, June 1996, pp. 56–58.

[Lysmer 70]   J. Lysmer, "Lumped Mass Method for Rayleigh Waves," *Bulletin, Seismological Society of America,* 1970, vol. 60, No. 1, pp. 89–104.

[Matzkanin 84]   G. A. Matzkanin, L. S. Fountain, and O. Tranbarger, "Nondestructive Evaluation of Infrastructure Conditions," *Infrastructure—Maintenance and Repair of Public Works, Annals of the New York Academy of Sciences,* vol. 431, Dec. 1984, pp. 268–303.

[Metzger 93]   D. S. Metzger, C. Barnes, and E. K. Miller, "Smart Roads," *Proceedings of the Smart Pavement Conference,* convened by ASTM Committee E-17 on Vehicle Pavement Systems, proceedings presented by Alliance for Transportation Research, Dallas, Tex., Dec. 1993, pp. 4–18.

[Nazarian 83]   S. Nazarian, K. H. Stokoe II, and W. R. Hudson, "Use of Spectral Analysis of Surface Waves Method for Determination of Moduli and Thickness of Pavement Systems," in *Transportation Research Record 930,* Transportation Research Board, National Research Council, Washington, D.C., 1983, pp. 38–45.

[NDT 96]   R. A. Miller, S. E. Swartz, and S. P. Shah, editors. *Proceedings, 2nd International Conference on Nondestructive Testing of Concrete in the Infrastructure,* Nashville, Tenn., Society for Experimental Mechanics, Inc., 1996.

[NOAA 94]   "Climatological Data," National Oceanic and Atmospheric Administration, National Climatic Center, Asheville, N.C., 1994.

[O'Leary 96]   P. N. O'Leary, R. R. Sartor, Y. Fu, and J. T. DeWolf, "Nondestructive Testing of Columns in Reinforced Concrete Bridge," *Proceedings, 2nd International Conference on Nondestructive Testing of Concrete in the Infrastructure,* Nashville, Tenn., 1996, pp. 222–229.

[Olson 96]   *Condition Assessment of Buildings and Bridges,* equipment brochure, Olson Engineering Inc,, 1996.

[Paterson 90]   W. D. O. Paterson and T. Scullion, "Information Systems for Road Management: Draft Guidelines on System Design and Data Issues," *Technical Paper INU77,* Infrastructure and Urban Development Department, The World Bank, Washington D.C., 1990.

[Pulse 94]   Pulse Radar Inc., *Ground Penetrating Radar Surveys, MDOT Study— US78 PCC Pavement,* report prepared for The University of Mississippi/ Mississippi Department of Transportation, May 1994.

[Schickert 96]   M. Schickert, "The Use of Ultrasonic A-Scan and B-Scan and SAFT Techniques for Testing Concrete Elements," *Proceedings, 2nd International Conference on Nondestructive Testing of Concrete in the Infrastructure,* Nashville, Tenn., 1996, pp. 135–142.

[Shahin 90]   M. Y. Shahin and J. A. Walter, "Pavement Maintenance Management for Roads and Streets Using the PAVER System," *USACERL Technical Report M-90/05,* Champaign, Ill., 1990.

[SHRP 89a]   "SHRP-LTPP Guide to Field Sampling and Handling," *SHRP 5021,* Strategic Highway Research Program, National Research Council, Washington, D.C., 1992.

[SHRP 89b]   "SHRP-LTPP Laboratory Guide for Testing Pavement Samples," *SHRP 5025,* Strategic Highway Research Program, National Research Council, Washington, D.C., 1992.

[SHRP 93]   "Distress Identification Manual for the Long-Term Pavement Performance Project," *Report SHRP-P-338,* Strategic Highway Research Program, National Research Council, Washington, D.C., 1993.

[Su 96]   M. B. Su, and Y. J. Chen, "Multiple Reflection of Metallic Time Domain Reflectometry," *Proceedings, 2nd International Conference on Nondestructive Testing of Concrete in the Infrastructure,* Nashville, Tenn., 1996, pp. 60–65.

[TRB 95]   L. D. Olson, F. Jalinoos, and M. F. Aouad, "Determination of Unknown Subsurface Bridge Foundations," *NCHRP Report 21-5,* Transportation Research Board, Washington, D.C., 1995.

[Uddin 83]   W. Uddin, S. Nazarian, W. R. Hudson, A. H. Meyer, and K. H. Stokoe II, "Investigations into Dynaflect Deflections in Relation to Location/Temperature Parameters and in Situ Material Characterization of Rigid Pavements," *CTR Report 256-5,* Center for Transportation Research, The University of Texas at Austin, 1983.

[Uddin 94]   W. Uddin, and W. R. Hudson, "Evaluation of NDT Equipment for Measuring Voids under Concrete Pavements," *ASTM STP 1198, 2nd International Symposium on Nondestructive Testing and Backcalculation of Moduli,* vol. 2, June 1994, pp. 488–502.

[Uddin 95a]   W. Uddin, "Pavement Material Property Databases for Pavement Management Applications," *Computerization and Networking of Materials Databases,* vol. 4, *ASTM STP 1257,* Philadelphia, Pa., 1995, pp. 96–109.

[Uddin 95b]   W. Uddin and V. Torres-Verdin, "Structural Capacity and User Cost Analyses for Road Investment Planning," *Proceedings, 2nd International Conference on Roads and Road Transport (ICORT-95),* Delhi, India, Dec. 1995.

[Uddin 96a]   W. Uddin, R. M. Hackett, P. Noppakunwijai, and Z. Pan, "Three-Dimensional Finite-Element Simulation of FWD Loading on Pavement Systems," *Proceedings, 24th International Air Transportation Conference,* Louisville, Ky., 1996, pp. 284–294.

[Uddin 96b]   W. Uddin, R. M. Hackett, P. Noppakunwijai, and T. Chung, "Nondestructive Evaluation and in Situ Material Characterization of Jointed Concrete Pavement Systems," *Proceedings, 2nd International Conference on Nondestructive Testing of Concrete in the Infrastructure,* Nashville, Tenn., 1996, pp. 242–249.

[Urban 84a]   The Urban Institute, *Guides to Managing Urban Capital,* vol. I: A Summary, H. P. Harty and G. E. Peterson (eds.), Washington, D.C., 1984.

[Urban 84b]   S. R. Goodwin and A. E. Peterson, *Guides to Managing Urban Capital Series,* vol. 2: *Guide to Assessing Capital Stock Condition,* The Urban Institute, Washington, D.C., 1984.

[Urban 84c]   H. P. Harty and B. G. Steinthal, *Guides to Managing Urban Capital Series,* vol. 4: *Guide to Selecting Maintenance Strategies for Capital Facilities,* The Urban Institute, Washington, D.C., 1984.

Uses of Monitoring Data and Examples of In-Service Evaluation

In-service evaluation of infrastructure facilities by nondestructive testing methods has become a popular approach to estimating

1. The physical and structural condition of aging structures
2. The effects of environmental and corrosion stresses
3. Post-evaluation of structural integrity after natural disasters such as floods and earthquakes, and accidents such as vehicular collisions or fire
4. Seismic safety and retrofitting needs

The available sensors and nondestructive evaluation (NDE) technologies were briefly reviewed in Table 6.4. Some examples of in-service evaluation for selected infrastructure facilities are presented in the following sections:

- Road and airport pavements
- Railroad tracks
- Bridges
- Water pipelines
- Gas pipelines
- Buildings

7.1   In-Service Evaluation of Road and Airport Pavements

Pavement management systems (PMS) for roads and highways have become established over the last three decades, and are supported by the Federal Highway Administration (FHWA) [FHWA 89] and many state highway agencies. PMS has been implemented on airport pavements by the Federal Aviation Administration (FAA) [FAA 82]. The activities of these agencies, and pavement-evaluation technologies, are well documented by Haas and Hudson [Haas 78, Haas 94]. Therefore, only a limited discussion of road infrastructure is presented here. These principles and databases are also used to manage roadway signs, traffic control devices, and traffic incidents on roadways.

7.1.1   Pavement monitoring and evaluation

The evaluation of pavements involves monitoring of one or more of the following: distress, structural capacity, riding comfort, safety, and appearance. Table 7.1 lists several evaluation measures influencing pavement condition and overall performance. Not all of these are essential for pavement management. At the network level, roughness is the only measure used by some agencies, but at the project level, all major objective measurements are used by some agencies. Pavement condition deteriorates with time; therefore, periodic measurements are required to develop distress history.

TABLE 7.1   Concepts for In-Service Monitoring and Evaluation of Road Pavements

Monitoring	Evaluation
Longitudinal roughness	Serviceability
Surface distress and defects (cracking, deformation, patches, disintegration, surface defects)	Deterioration, overall composite index, and maintenance needs
Deflection testing	Material properties and structural capacity
Skid resistance or surface friction	Safety against skidding
Ride quality	User evaluation of overall pavement quality
Appearance	Aesthetics
Traffic	Performance and remaining life
Costs (construction, maintenance, user)	Unit-cost summaries for economic evaluation
Location reference, geometric and structure data, longitudinal and cross-fall deficiency, coring for layer thickness	Verification of inventory database, inputs for structural evaluation, safety against potential hydroplaning
Environment (climate, pavement temperature, drainage, water below surface, freeze/thaw)	Material degradation, distress and defects progression, structural integrity, performance

7.1.2   Pavement nondestructive evaluation

The major type of pavement nondestructive evaluation often measured and used to evaluate structural capacity and residual life is deflection testing. The Benkelman Deflection Beam was developed at the WASHO Road Test in the mid-1950s and still remains in wide use around the world. For over two decades, most pavement rehabilitation design procedures have been based on Benkelman Beam deflections. The ever-growing demand for faster, easier to use, and more automated nondestructive testing (NDT) equipment has resulted in the development of vibratory loading devices like the Dynaflect and Road Rater, and the impact loading falling weight deflectometer (FWD). Commercially available dynamic deflection equipment measures and records the dynamic response of a road or airport pavement subjected to steady-state vibratory or transient loading forces. Examples are Dynaflect, Road Rater, and FWD. Improvements have been made to automate the testing sequence and on-board data processing. Table 6.5 shows guidelines for the deflection data sampling rate for roads.

Deflection test data are also used to estimate in situ material characterization, and to evaluate the structural integrity of concrete pavements [Hudson 87a]. The in situ material properties and structural capacity are calculated from the dynamic deflection data using static load analytical procedures and empirical performance relationships [Uddin 86], as shown in Figure 7.1. A significant step in this area is the development of three-dimensional finite-element dynamic analysis procedures [for example, Uddin 96a] for interpretation of the dynamic deflection data.

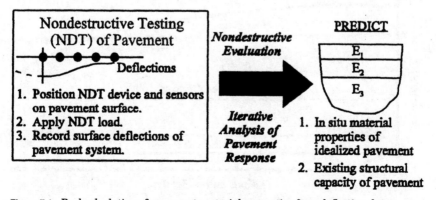

Figure 7.1 Backcalculation of pavement material properties from deflection data.

7.1.3 The serviceability and performance concept

Serviceability is the ability of a specific section of pavement to serve traffic in its existing condition. Present Serviceability Index (PSI) can be calculated from roughness data on a scale of 5 to 0. Performance is a measure of the accumulated service provided by a facility, i.e., the adequacy with which a pavement fulfills its purpose.

The evaluation of pavement performance involves a study of the functional behavior of a section or length of pavement. For a functional or performance analysis, information is needed on the history of the riding quality of the pavement section for the time period chosen and the traffic during that time. This can be determined by periodic observations or measurements of the pavement riding quality, coupled with records of traffic history and time. It is this history of deterioration of the ride quality or serviceability provided to the user that defines pavement performance (see Figure 7.2).

Until a measure of pavement serviceability was developed in conjunction with the AASHO Road Test [HRB 62], inadequate attention was paid to the evaluation of pavement performance. A pavement was considered to be either satisfactory or unsatisfactory (i.e., in need of repair or replacement). Pavement design technology did not directly consider performance. Design engineers have varied widely in their concepts of desirable performance. For example, at one extreme an engineer asked to design a pavement for a certain expected traffic level for 20 years might consider the job properly done if little or no cracking occurred during the design period. On the other hand, a second designer might be satisfied if the pavement had reached a totally unacceptable level of serviceability at the end of the design period.

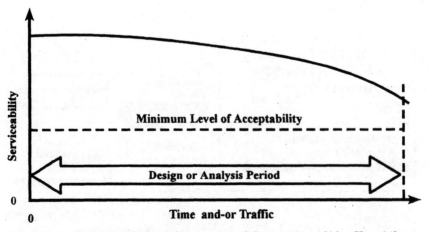

Figure 7.2 Deterioration of ride quality or serviceability over time. [After Haas 94].

7.1.4 Relating roughness to serviceability

The primary use of objective roughness measurements is to estimate the pavement serviceability, which is a subjective (i.e., user) rating of pavement ride quality. The first and most widely used method for this purpose was the Present Serviceability Index (PSI), at the AASHO Road Test [HRB 62]. The original functional form of the PSI equation is:

$$PSI = C + (A_1R_1 + ...) + (B_1D_1 + B_2D_2 + ...) \pm e \qquad (7.1)$$

where C = coefficient (5.03 for flexible pavements and 5.41 for rigid pavements);

A_1 = coefficient (-1.91 and -1.80 for flexible and rigid, respectively);

R_1 = function of profile roughness [$\log(1 + SV)$], where SV = mean slope variance obtained from the CHLOE profilometer;

B_1 = coefficient (-1.38 for flexible and 0 for rigid);

D_1 = function of surface rutting (RD), where RD = mean rut depth as measured by simple rut-depth indicator;

B_2 = coefficient (-0.01 for flexible and -0.09 for rigid)

D_2 = function of surface deterioration ($C + P$), where $C + P$ = amount of cracking and patching, determined by procedures developed at the AASHO Road Test; and

e = error term.

Given this general form of equation, it is necessary to determine the coefficients for a particular set of input variables. This was done at the AASHO Road Test for several sets of variables [Carey 60]. It is important to understand that the result is a best-fit equation based on all observed data used in the equation. Other variables were candidates for inclusion in the equation, but they added no significance in predicting PSI.

Additionally, a regression equation is not a causative relationship, and covariance between terms can account for very small coefficients on a variable that alone is only slightly less well correlated with the dependent variable. For example, if the observed roughness in a pavement is due to cracking, the two factors are correlated. Consequently, once the roughness term is included in the equation, little variation remains to be explained by adding the cracking terms and thus the coefficient is small. This does not indicate a lack of concern for cracking, but if users of the equation alter it arbitrarily because they intuitively "feel" that cracking is more important, then erroneous and unpredictable results will occur.

7.1.5 Evolution of the serviceability-roughness concept

Any change in measurement methods or units will result in a modified equation. This can be done either by performing an entirely new regression if all data are available, or by comparing the old measurement to the new and making an appropriate substitution in the equation. For example, at the AASHO Road Test [HRB 62] the BPR roughometer output, R, in inches per mile, was correlated to slope variance, SV, and substituted into eq. (7.1). The resulting equation for rigid pavements was:

$$PSI = 5.41 - 1.80 \, \text{Log}(0.40R - 30) - 0.09\sqrt{(C + P)} \qquad (7.2)$$

It must be emphasized that any PSI model, such as that represented by eq. (7.1), is not an end in itself. Carey and Irick made this quite clear in pointing out that it is intended only to predict PSR to a satisfactory approximation [Carey 60].

7.1.6 Distress monitoring and evaluation

Pavement distress is observable deterioration or damage in the pavement. Distress data attributes include cracking, permanent deformation, and disintegration caused by load, age, environment, material deficiency and degradation, inadequate pavement design or poor construction quality, and interaction among these factors. Because maintenance may have been performed on some of the distress, the evidence of this maintenance in the form of patches and sealed areas is a distress and also should be monitored. Distress data (type, severity, and extent) are useful measures for selecting a proper M,R&R alternative. As discussed in a later section, the distress data collected on a pavement section can be combined in single numeric statistic, such as the Pavement Condition Index (PCI) on a scale of 100 to 0 [Shahin 90]. Figure 7.3 shows an example of the PCI and PSI scales related to overall pavement condition and maintenance needs [OECD 87].

7.1.7 Safety evaluation

Safety may be measured by skid resistance or friction measurements or in an empirical fashion through determination of those locations with high accident rates. However, safety hazards may not be due to pavement-related factors, but could, for example, indicate an alignment problem. Such factors may be included in the roadway-management system, at the discretion of the agency involved. The typical current practice is to use skid resistance as indicated by pavement surface friction as the primary measure of pavement-related safety.

Figure 7.3 Significance of PCI and PSI for pavement evaluation. [after OECD 87].

7.1.8 Selection of pavement evaluation technologies

The selection of a specific type of technology for each data category depends on the data quality needed, data collection speed/productivity, and available budgets. The equipment and methods can be grouped into three classes [Hudson 87a]: (1) visual distress and rutting (detailed slow walk survey or high-speed photographic and video survey); (2) stand-alone automated equipment (for example, dedicated roughness equipment like Maysmeter or Bump Integrator), deflection devices like Dynaflect or falling weight deflectometer, skid-resistance measuring equipment; and (3) multipurpose automated vehicles, high-speed equipment used to collect distress, roughness, rutting and other condition/video data during a single pass. Some multipurpose equipment is listed in Table 5.5 in Chapter 5.

Multifunction automated evaluation equipment currently offers the desired productivity, cost effectiveness, and electronic data-processing capabilities for both network-level and project-level pavement management applications, and they are still being improved upon. [Haas 94]

7.1.9 Airport pavement evaluation

Pavement distress is often used as the condition measure of airfield pavements. Deflection tests are also used for nondestructive evaluation. Factors contributing to the use of distress in preference to roughness for the evaluation of airfield pavement include [Haas 94]:

1. Relatively limited surface area of an airfield permits the collection of more detailed distress data and therefore an improved ability to rate pavement quality

2. Support of military and FAA for the PAVER method to evaluate PCI.

3. Lack of development of a serviceability concept comparable to the highway application

4. Damage potential if loose pavement pieces are injected into an aircraft engine (Foreign Object Damage, FOD)

These factors, however, do not mean roughness should be excluded from indicators of airfield pavement performance measures. In fact, rough runway pavements contribute to the fatigue of aircraft structural members. Profiler devices are required to meet the critical roughness criteria for airfields to simulate the aircraft landing and take-off patterns. The recently developed aircraft ride quality index (RQI), based on aircraft landing or take-off simulations on pavement longitudinal profile measured by laser surveys, provides an improved procedure for airfield pavement roughness evaluation. The RQI is calculated from the vertical acceleration experienced at the pilot seat and the aircraft center of gravity. If the acceleration exceeds 0.4 g, the ride quality is in the area of human discomfort [Gerardi 96].

7.2 In-Service Evaluation of Railroad Tracks

An effective track-management system requires periodic monitoring and evaluation of track structure. As mentioned earlier, Union Pacific has embarked on high-tech video and GIS technology for in-service inspection [Gerard 91]. Inventory elements include details of gauge width, type of ballast, type of joints, date of construction, and so forth.

7.2.1 Track quality index (TQI)

The evolution of track condition assessment is summarized by Fazio and Prybella [Fazio 80]. In general terms, track quality can be defined as "the ability of the track structure to meet its functional requirements." According to this definition, the type of operations supported by the track will influence the selection of the parameters used to measure quality. Although many railroads have in-house estimators of track quality—such as the "condition index" used by Conrail (Consolidated Rail Corporation)—existing estimators either lack universality or require the collection of excessive amounts of field data to calculate the parameter for a given track segment.

A measure of track quality, called track quality index (TQI), has been developed in a Federal Railroad Administration (FRA) sponsored research project. The TQI statistic can be derived primarily from data collected by automatic track geometry cars. Parameters that can be measured by these cars include gauge, cross level, warp (rate of change of cross level), and alignment. The FRA-owned cars are capable of recording measurements every 0.7 m (2 ft) at speeds up to 242 km/h (150 mph) and carry on-board computers so that a computer tape of the collected data, in addition to the strip chart supplied to maintenance-of-way personnel, is available at the conclusion of the survey. Each of these parameters of track geometry has a unique distribution over a section of track that can be used to generate various summary statistics. The TQI parameters also can include other track data, such as bad-tie counts, that are collected by manual inspections. This might be avoided by developing the capability to estimate track modulus (the ability of the track to resist deflection under load) by using automated track geometry cars.

Operating and loading conditions will affect the required level of TQI. For instance, different TQIs would be applicable to tracks that support 80 km/h (50 mph) freight traffic and tracks that support 160 km/h (100 mph) Metroliner service. The FRA-Conrail project focused on operating conditions that are common to many North American freight railroads: mixed freight trains of 50 to 100 cars operating on conventional track (wood tie and stone ballast) at speeds of 64-80 km/h (40–50 mph). In Europe, the Netherlands Office of Research and Experiments (ORE) also has attempted to develop an easily measurable quantifier of track condition based on automatic track geometry cars [Fazio 80].

7.2.2 Railroad track condition evaluation

The Construction Engineering Research Laboratory (CERL) has developed a track maintenance management program, RAILER, for inventory, inspection, evaluation, and maintenance/repair cost analysis. Defects are rated according to five pre-established condition levels: no defects, no restrictions, 10 mph, 5 mph, and no operation. It also allows automated track geometry data input. The maintenance and repair needs are assessed by comparing the visually collected data and rail-flaw inspection information with the railroad track standards [Uzarski 88].

7.3   Evaluation of Bridges

The first national bridge management study [Hudson 87] included an in-depth review of the existing NBIS procedures of bridge-structure inventory and condition appraisal (SI&A) standards, identified several improvements in these procedures to enhance in-service evaluation for meaningful M,R&R programming, and recommended a model bridge-management system (BMS).

7.3.1   Federal SI&A data and limitations

The bridge condition variables, included in the federal SI&A standards, generally exist in the form of a single severity rating for each of the major components of a bridge. These ratings are used during visual bridge inspections. Table 7.2 lists these major components and the detailed condition variables recommended by the BMS study.

The following limitations in SI&A data were identified in NCHRP Report 300 [Hudson 87]:

TABLE 7.2   Summary of Bridge Condition Variables Recommended for an Ideal BMS

	BMS Recommended Condition Variables	
Federal SI&A rating group	Number	Examples
Roadway condition rating	8	Deck, wearing surface, joints, drainage, curb/sidewalk/parapet, median barrier, railing, delineation
Superstructure condition rating	9	Main members and connections, floor members and connections, secondary members and connections, expansion and fixed bearings, steel protective coating
Substructure rating	7	Abutments, intermediate supports, collision protection system, steel protective coating, retaining walls, culverts, concrete protective system
Channel and channel protection rating	5	Banks, bed, rip rap, dikes and jetties, substructure foundation erosion
Approaches rating	6	Embankments, pavements, relief joints, drainage, guard fence, delineation markers
Estimated remaining life	1	Estimated remaining life
Date and other information inspection	5	Dates of last inspection and unusual inspection features, frequency and date of last unusual inspection, inspector

- Lack of distress extent and type data.
- No flagging mechanism for urgent actions.
- Ratings do not reflect maintenance needs (type and quantity); for example, a deteriorated deck condition may be a costly major maintenance item, but it has little influence on the structural condition severity ratings.
- Low precision and reliability of the ratings and their associated definitions, especially in gray areas of intermediate condition ratings (it is difficult to differentiate between needs for minor maintenance, major maintenance, and minor rehabilitation).
- Lack of objective instrumentation that is reliable and measures valid parameters.
- Lack of maintenance items on SI&A forms (for example, painting is not considered in the inspection).
- Field inspection is geared to rehabilitation and replacement (structural, adequacy, safety considerations), and not to minor or preventive maintenance.
- Inadequate maintenance data, cost-effectiveness data, and inadequate overall condition index for network-level prioritization.

The Federal Sufficiency Rating (SR) is based on the SI&A ratings and used to determine the eligibility of bridges for funding under the Federal Highway Bridge Replacement and Rehabilitation Program (HBRRP). Other ratings are also used—load capacity in terms of Inventory Rating (IR) and Operating Rating (OR). The SI&A ratings are made on a scale of 9 (excellent condition) to 0 (worst condition). SR ranges from 100 to 0, and is calculated as follows [Hudson 87b]:

$$SR = S1 + S2 + S3 - S4 \tag{7.3}$$

The variables $S1$, $S2$, $S3$, and $S4$ are defined below:

1. $S1$ = Structural adequacy and safety (55 maximum, 0 minimum)

$$S1 = 55 - (A + I) \tag{7.4}$$

where A is used to deduct values for deterioration, based on superstructure rating and substructure rating, and I is used for reduction for load capacity

$$I = (36 - AIT)^{1.5} \times 0.2778 \tag{7.5}$$

where AIT is adjusted inventory rating.

2. $S2$ = Serviceability and functional obsolescence (30 maximum, 0 minimum)

$$S2 = 30 - [J + (G + H) + I] \qquad (7.6)$$

where J shall not exceed 13; $(G + H)$ shall not exceed 15; and I shall not exceed 2. J is a linear function of roadway condition, structural condition, roadway geometry, and underclearance; G and H are functions of roadway width insufficiency; and I is a function of vertical clearance insufficiency.

3. $S3$ = Essentially (15 maximum, 0 minimum)

$$S3 = 15 - (A + B) \qquad (7.7)$$

where A and B are reductions for public and military uses, respectively.

4. $S4$ = Special reductions (to be used only when $S_1 + S_2 + S_3$ are equal to or more than 50)

$$S4 = A + B + C \qquad (7.8)$$

where A is detour-length reduction; B is structure-type reduction; and C is highway-safety-feature reduction. A and B are reductions for public and military uses, respectively. Maximum value of A, B, or C is 5.

7.3.2 Survey of bridge condition evaluation methods

The AASHTO manual for condition evaluation of bridges is the result of cooperative research conducted under NCHRP project 12-23 in cooperation with the Federal Highway Administration [AASHTO 94]. The manual is intended to establish standard inspection procedures and load-rating practices. Tables 7.3, 7.4, and 7.5 are based on the AASHTO manual and provide a comparison of the capabilities of the test methods for the evaluation of concrete, steel, and timber structures.

At the 20th World Road Congress of the Permanent International Association of Road Congresses (PIARC) in Montreal in 1995, an international survey of the techniques for assessing in-service condition of bridges was reported [PIARC 95]. The following information is based on a summary of that survey information from Europe, the United States (New Jersey), and Japan.

Inspection intervals vary according to the importance of the bridge, the age, and type of structure (steel or concrete), the degree of damage, and traffic conditions. The number of data items for bridge inspections and condition assessment range from 20 (France) to 250 (United Kingdom). Visual inspections and equipment-aided inspections are carried out in most countries. Inspection walkways or mov-

TABLE 7.3 Capability of Evaluation Techniques for Concrete Structures [AASHTO 94]

Method Based on	Capability of Defect Detection					
	Cracking	Scaling	Corrosion	Wear and Abrasion	Chemical Attack	Voids in Grout
Strength	Unsuitable	Unsuitable	Poor	Unsuitable	Poor	Unsuitable
Sonic	Fair	Unsuitable	Good²	Unsuitable	Unsuitable	Unsuitable
Ultrasonic	Good	Unsuitable	Fair	Unsuitable	Poor	Unsuitable
Magnetic	Unsuitable	Unsuitable	Fair	Unsuitable	Unsuitable	Unsuitable
Electrical	Unsuitable	Unsuitable	Good	Unsuitable	Unsuitable	Unsuitable
Nuclear	Unsuitable	Unsuitable	Fair	Unsuitable	Unsuitable	Unsuitable
Thermography	Unsuitable	Good¹	Good²	Unsuitable	Unsuitable	Unsuitable
Radar	Unsuitable	Good¹	Good²	Unsuitable	Unsuitable	Unsuitable
Radiography	Fair	Unsuitable	Fair	Unsuitable	Unsuitable	Fair

[1]Beneath bituminous surfacing.
[2]Detects delaminations.

TABLE 7.4 Capability of Evaluation Techniques for Steel Structures [AASHTO 94]

Method Based on	Cracks				Internal Voids	Welds A	Thick B	Stress C	Blist D	Corr E
	Minor Surface	Deeper Surface	Internal	Fatigue						
Radiography	N¹	Fair²	Fair²	Poor	Good	Good	Fair	Fair	Poor	Good
Magnetic particle—wet	Good	Good	N	Good	N	N	N	Good	N	N
Magnetic particle—dry	Fair	Good	N	Good	N	N	N	Fair	N	Poor
Eddy current	Fair	Good	N	N	N	Poor	Poor	N	N	N
Dye penetration	Fair	Good	N	Good	N	N	N	Good	N	Fair
Ultrasonic³	Poor	Good	Good	Good	Good	Fair	Good	Fair	Fair	Poor

Note: A = Porosity and slag in welds; B = thickness; C = stress corrosion; D = blistering; E = corrosion pits.
[1]N = Not suitable.
[2]If beam is parallel to cracks.
[3]Capability varies with equipment and operating mode

able platforms are installed as permanent inspection facilities in all countries. Boom vehicles and over-the-fence operation vehicles are also used. The survey included a review of primary damage mechanisms and inspection techniques in the following categories of bridge components: steel members, concrete members, piers, abutments,

TABLE 7.5   Capability of Evaluation Techniques for Timber Structures [AASHTO 94]

Method Based on	Surface Decay and Rot	Internal Decay and Voids	Weathering	Chemical Attack	Abrasion and Wear
		Capability of Defect Detection			
Penetration	Good	Good	Fair	Fair	Unsuitable
Electrical	Fair	Fair	Unsuitable	Unsuitable	Unsuitable
Ultrasonic	Unsuitable	Good	Good	Unsuitable	Unsuitable

foundations, and ancillary equipment. Determination of causes of damage is an important part of visual inspection and testing. Types of condition deterioration and inspection methods for these components are summarized in Table 7.6

7.4   Evaluation of Water Pipelines

The following section focuses on leakage, corrosion, and structural deterioration of water mains. O'Day groups water-distribution-system deterioration problems into the following areas [O'Day 84]:

- Water-quality problems related to tuberculation and internal pipe corrosion.

- Low-pressure and high head-loss problems due to main tuberculation.

- Distribution system leakage due to joint leaks, main breaks, and service leaks.

- Main breaks due to main deterioration from internal and/or external corrosion.

- Valves and hydrants inoperable due to neglect and/or unavailable replacement parts.

The forces acting on the mains are: (1) internal pressure; (2) external pressure from surrounding earth, truck load, and frost in cold climates; (3) bending stress if the pipe is not uniformly supported; and (4) thermal stresses if the main is restricted from expansion and contraction. Internal, external, and electrolytic corrosion of pipe material leads to the loss of wall thickness. Soil resistivity and water content play a major role in corrosion rates. Leakage from water mains tends to increase corrosion deterioration. Figure 7.4 shows the factors that affect the forces on the mains and those that affect the loss of pipe wall [O'Day 84].

TABLE 7.6   Summary of Bridge Condition Evaluation Methods Based on a Survey
of 17 Countries [PIARC 95]

Bridge Component	Deterioration and Damage	Inspection Method
Steel members	(1) Paint-film aging and premature damage due to rust and corrosion	(1) Visual and photo inspection; tape pull-off and adhesion tests
	(2) Loss of joint bolts or rivets and weld damage; cracks due to fatigue and corrosion; buckling; brittle fracture	(2) Detection of abnormal noise or vibration; radiograph and ultrasonic testing; vibration tests
Concrete members	(1) Reinforcement corrosion and cracking due to salt damage	(1) Visual inspection for cracks and all other observable damage
	(2) Surface cracking and corrosion due to alkali-silica reaction (ASR)	(2) Visual inspection; concrete strength test; gel examination, chemical ASR tests; X-ray analysis (not common) of mortar bar and aggregate
	(3) Reinforcement corrosion and cracking due to concrete carbonation	(3) Phenolphthalein method to detect carbonation (most common method)
	(4) Prestress steel corrosion due to deteriorated duct grouping	(4) Drilling and visual inspection; nondestructive testing by x-ray, ultrasonic, acoustic emission
	(5) Concrete slab fatigue cracking, spalling, delamination, and damage	(5) Visual inspection; concrete strength testing; corrosion condition evaluation
	(6) Cracks due to excessive loading (near the center of span, at quarter point of span, the periphery of supports)	(6) Visual inspection; ultrasonic testing or coring to check crack depth

Note: Concrete strength by core testing, Schmidt hammer, ultrasonic testing (See Chapter 6 for new and innovative NDE evaluation methods.)

Piers, abutments, foundations	(1) Deterioration of steel and concrete members	(1) Visual inspection and other methods listed above
	(2) Damage due to accidents (impact of ships, floating logs, vehicles) or natural disasters (earthquake, floods)	(2) Site and incident specific inspection; seismic evaluation and liquefaction potential
(Critical for overall safety of the bridge)	(3) Deformation and damage due to ground changes (erosion and scouring, settlement. Tilting, substructure problems)	(3) Frequent monitoring of: erosion and scouring by using poles and drivers and ultrasonic testing; settlement by survey, depression meter, ultrasonic testing; tilting by survey, tilt meter; substructure inspection by test excavation and machine boring

TABLE 7.6 Summary of Bridge Condition Evaluation Methods Based on a Survey of 17 Countries [PIARC 95] (*Continued*)

Bridge Component	Deterioration and Damage	Inspection Method
Ancillary equipment	(1) Damage to bearings (restriction to rotation and movement, lack of mobility, rust and corrosion, deformation of bearing seat or anchor bolts or sole plates)	(1) Mostly checked by visual inspections; mobility gages
	(2) Damage to expansion joints (bumps between traveling surfaces, abnormalities in joint openings, loosening of bolts, fracture of fingers, water leaks, abnormal noise and vibrations)	(2) Mostly checked by visual inspection; mobility checks; anchor-bolt torque test; hammer test

Figure 7.4 Conceptual model of water-main structural condition. [after O'Day 84].

Leakage and break frequency are important evaluation parameters for water mains. Leakage monitoring is performed by actual flow test, sonic listening. Nondestructive and noncontact testing by infrared thermography is another method to locate areas of potential breaks and leakage.

The following five levels of analysis of leakage and break data are routinely collected by utility companies [O'Day 84]:

1. Analysis of break trends over time to determine the problem areas.

2. Analysis of break patterns by main type (considering material type, diameter, age, joint type, importance of the area, and other urgency factors) to define the main types and areas that experience the greatest rate of failure.

3. Analysis of break patterns by break type to find out the break type that is linked to main diameter, age, material, and location. This should be recorded at the time of the repair, to know the cause of structural deterioration and failure. Circular breaks are generally common on small-diameter (equal to or less than 25.4 mm or 10 in) pipes, and longitudinal breaks are frequently recorded on large-diameter pipes.

Break Type	Cause of Structural Failure
Circumferential	Thermal contraction, bending
Longitudinal	Excessive crushing load
Hole	Internal pressure
Bell crack	Thermal contraction, joint material expansion

4. Visual inspection of break samples is used to determine the source, extent, and depth of corrosion.

5. Microscopic examination of break samples can identify the corrosion type by the microstructure analysis of the metal.

7.5 Evaluation of Gas Pipelines

Natural gas is transported under high pressure through steel pipelines with protective coating. The Natural Gas Pipeline Safety Act of 1968, the Hazardous Liquid Pipeline Act of 1979, and the Pipeline Safety Reauthorization Act of 1988 require the U.S. Department of Transportation (DOT) to develop and enforce safety regulations [GAO 92]. There are more than 2.6 million km (1.6 million mi) of natural gas pipelines in the United States, including 427,000 km (267,000 mi) of interstate natural gas pipelines. According to a report by the General Accounting Office (GAO), the safety of these pipelines is becoming a concern as they age; for example, 9 percent of the network was constructed before 1940, 10 percent before 1950, and most were constructed during the 1950s and 1960s. The following discussion is based on the GAO report.

From 1985 to 1991, 1726 natural-gas pipeline accidents involving 131 fatalities and 634 injuries were reported. The leading cause of pipeline failure is outside-force accidental damage due to outside force such as a third-party excavation; the second cause is internal and external corrosion. The pipeline must be protected during its service life from damage and degradation, and this is generally accomplished by external corrosion controls and inspection techniques. The pipeline inspection and evaluation methods include:

- Visual inspection, line walking, line walking with a flame detector, portable ultrasonic sensor to check the structural integrity, and use of light aircraft or helicopter to check for evidence of leaking, such as dying vegetation, water bubbles, and cavities in the ground
- Hydrostatic pressure testing of the pipeline by forcing water into a segment of pipeline and monitoring its resistance against rupture
- X-ray analysis to check pipe welds
- Placing an instrumented device, such as smart pig, inside the pipe to record flaws as the instrument is transported through the pipe by the natural gas

Smart pigs have been used since the 1960s. They carry ultrasonic or magnetic flux leakage-measuring instruments to monitor internal and external corrosion and other flaws. This technology is very cost-effective and does not require a break in service for hydrostatic pressure testing. The GAO report provides details about conventional and smart-pig technologies for inspection and testing of gas pipelines [GAO 92].

7.6   Evaluation of Buildings

Buildings are the most vital infrastructure assets for proper functioning of public life at private, commercial, and public levels. Proper functioning of buildings requires inspections and maintenance of structural aspects, as well as heating and cooling, ventilation, water and drainage, emergency and security systems, horticulture, parking, and other nonstructural components. Unfortunately, as-built records of building construction and M,R&R actions generally are not adequately preserved and accessible.

7.6.1   Example of building condition rating and evaluation

The inspection and building assessment approach also depends upon the insight and management approach of owners, operators, and regulating agencies. Table 7.7 shows a public facility rating form used by the U.S. Postal Service for its decision analysis [after Urban 84c].

Building components have wide ranges of life expectancy, depending upon the material, design and construction practices, soil and environment, and occurrences of natural hazards. By far, fire has been the most common type of accidental hazard to ordinary buildings.

TABLE 7.7   Example of U.S. Postal Service Public Facility Rating Form [after Urban 84c]

General Mail Facility—Working Conditions Summary, Lancaster, Pennsylvania										
	Condition			Area of Interest						
Components	A	P	I	H	M	Sa	Se	E	C	Remarks
Air conditioning			X	X				X		Window units only
Electrical		X			X					Octopus system difficult to maintain
Heating	X									Recent major modifications
Lighting			X			X		X		Office levels of 20–30 foot candles
Plumbing		X						X		Reduced flows
Offices		X						X		Lighting, HVAC
Lobby		X				X			X	Lighting levels low
Lockers			X	X		X	X	X		All-around deficiencies
Lunch room				X					X	All-around deficiencies
Rest rooms		X		X				X		Deficient by criteria
Support area	X									—
Work room		X				X		X		Poor lighting levels
Platform			X			X		X		Canopy repairs required
Parking area			X					X	X	No carrier parking
Maneuvering area			X							Lack of sufficient depth
Structure	X									—
Miscellaneous	X									—

Note: A = adequate; P = poor; I = inadequate; H = health; M = maintenance; Sa = safety; Se = security; E = employee accommodation; and C = customer accommodation.

7.6.2 Postearthquake evaluation

Earthquakes are important natural disasters that are difficult to predict. Figure 7.5 shows areas of significant earthquakes in the United States. The plot shows contours of effective peak acceleration (expressed as a decimal fraction of gravity, g) that might be expected (with odds of one in ten) to be exceeded during a 50-year period [FHWA 81, Buckle 87].

Postearthquake safety evaluation, posting, and condemnation of buildings are now desired by federal and state emergency management agencies, and staff members of concerned agencies are being trained in earthquake-prone areas. This can be a mammoth task for a large urban area; for example, San Francisco has over 150,000 buildings, 30 building inspectors, and 15 civil planning and structural

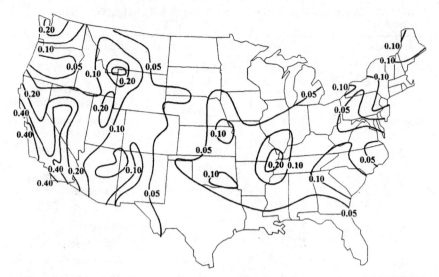

Figure 7.5 Areas of significant earthquakes in the United States. [adapted from FHWA 81, Buckle 87].

engineers working with the government agency. If 10 percent of buildings need inspection, then the buildings/inspector ratio is 300 to 1 [ATC 93]. The ATC training manual, prepared with funding from the Federal Emergency Management Agency (FEMA), outlines some of the key concerns, as summarized below:

- *Principal safety concerns* (collapse, falling hazards, geotechnical hazards, hazardous materials, field safety)

- *Posting system* (timely, consistent, visible, authority of jurisdiction); Inspected (green)—appears safe for lawful occupancy; Limited entry/restricted use (yellow)—Some restriction on use, controlled by tenant; Unsafe (red)—entry controlled by jurisdiction

- *Evaluation procedures* (windshield survey to assess scope of damage; rapid, to assess obvious and smaller buildings and recommend detailed evaluation where needed; detailed, to assess difficult and large buildings closely; and engineering, using consultants engaged by owner)

- *Evaluation steps* (for rapid and detailed evaluation)
  1. Examine entire outside of building.
  2. Examine ground for distress.
  3. Enter if safe and continue inspection.
  4. Discuss observations; evaluate by criteria.

5. Post building.
6. Inform occupants of hazards.

- *Detailed evaluation of essential facilities* (health-care facilities, police and fire stations, jails and detention centers, emergency operations centers, high-occupancy shelters)
- *Detailed evaluation criteria* (for hazardous condition)
  1. Vertical load capacity not significantly decreased
  1. Lateral load capacity not significantly decreased
  3. No falling or other hazards present
  4. No evidence of foundation damage or ground displacement
  5. Main exits are useable
  6. No other unsafe condition
- *Hazards of nonstructural elements*
  1. Parapets, chimneys, ornaments
  2. Cladding and glazing
  3. Partitions
  4. Suspended ceilings, raised floors
  5. Piping and duct work
  6. Equipment, furnishings
  7. Portable buildings
- *Geotechnical hazards* (Increased ground shaking, liquefaction, subsidence)

7.6.3 ASCE standards

The American Society of Civil Engineers (ASCE) Committee on Structural Condition Assessment and Rehabilitation of Buildings developed standard guidelines for structural condition assessment in 1990 [ASCE 90, Pillared 96] with the support of the Buildings and Fire Research Laboratory of the National Institute of Standards and Technology (NIST). The ASCE standard was the first coordinated effort to provide guidelines for assessing the structural condition of existing buildings constructed of concrete, masonry, metals, and wood. Due to potential costs, a multilevel approach is recommended, which starts with a preliminary assessment for initial structural adequacy and establishes the need and priority for detailed assessment. This is followed by a detailed assessment that recommends a course of action and alternatives.

Another ASCE subcommittee is preparing guidelines for condition assessment of building envelopes. Building envelope is defined as "the exterior surface of a building providing protection from weather." The proposed standards work, which was motivated by concern for public safety, outlines the following six envelope categories [Pillared 96]:

- Roof systems are designed to weatherproof and insulate the top surface of a building.

- Balcony systems are exterior platforms supported at the exterior walls of a building.

- Plaza deck systems are structures with a waterproof membrane applied to a horizontal surface that is intended to receive vehicular or pedestrian traffic.

- Bearing wall systems support gravity and vertical loads.

- Nonbearing wall systems are not intended to support vertical loads other than their own; for example, curtain and panel walls.

- Foundation wall system is the part of the building in contact with the soil. It is usually a load-bearing wall, often with waterproof coating and drainage components.

The ASCE Subcommittee on Seismic Rehabilitation of Buildings was established to assist in the consensus review of the seismic rehabilitation document ATC-28, prepared by the Applied Technology Council (ATC) of the Structural Engineers Association of California with funding from FEMA [Pillared 96].

References

[AASHTO 94] *Manual for Condition Evaluation of Bridges,* American Association of State Highway and Transportation Officials (AASHTO), Washington, D.C., 1994.

[ASCE 90] *ASCE 11-90, Standard Guidelines for the Structural Condition Assessment of Existing Buildings,* American Society of Civil Engineers, New York, 1990.

[ATC 93] Applied Technology Council, "Postearthquake Safety Evaluation of Buildings Training Manual," *ATC-20-T,* funded by the Federal Emergency Management Agency, Washington, D.C., 1993.

[Buckle 87] I. G. Buckle, R. L. Mayes, and M. R. Button, "Seismic Design and Retrofit Manual for Highway Bridges," *Report FHWA-IP-87-6,* Federal Highway Administration, McLean, Va., May 1987.

[Carey 60] W. N. Carey, and P. E. Irick, "The Pavement Serviceability—Performance Concept," *Highway Research Bulletin 250,* Highway Research Board, National Research Council, Washington, D.C., 1960.

[FAA 82] "Guidelines and Procedures for Maintenance of Airport Pavements," *Advisory Circular AC: 150/5380-6,* Federal Aviation Administration, Washington, D.C., 1982.

[Fazio 80] A. E. Fazio, and R. Prybella, "Development of an Analytical Approach to Track Maintenance Planning," in *Transportation Research Record 744,* Transportation Research Board, National Research Council, Washington, D.C., 1980, pp. 46–52.

[FHWA 81] "Seismic Design Guidelines for Highway Bridges," Applied Technology Council, *Report FHWA/RD-81/081,* Federal Highway Administration, Washington, D.C., 1981.

[FHWA 89]   "Pavement Management Systems, A National Perspective," *PAVEMENT Newsletter,* Federal Highway Administration, U.S. Department of Transportation, Issue 14, Spring 1989.

[GAO 92]   "Natural Gas Pipelines—Greater Use of Instrumented Inspection Technology Can Improve Safety," GAO Report to Congressional Committees, *Report No. GAO/RCED-92-237,* United States General Accounting Office, Washington, D.C., 1992.

[Gerard 91]   S. V. Gerard, "UP Links Video and Graphic Images for Greater Clarity," *Railway Track and Structures,* April 1991.

[Gerardi 96]   T. Gerardi, "The Importance of Maintaining Smooth Airport Pavements," *Proceedings, 24th International Air Transportation Conference,* Louisville, Ky., pp. 295–305.

[Haas 78]   R. C. G. Haas and W. R. Hudson, *Pavement Management Systems,* McGraw-Hill, New York, 1978.

[Haas 94]   R. Haas, W. R. Hudson, and J. P. Zaniewski, *Modern Pavement Management,* Krieger Publishing Company, Malabar, Fla., 1994.

[HRB 62]   "The AASHO Road Test, Report 5—Pavement Research," *Special Report 61-E,* Highway Research Board, National Research Council, Washington, D.C., 1962.

[Hudson 87]   S. W. Hudson, R. F. Carmichael III, L. O. Moser, W. R. Hudson, and W. J. Wilkes, "Bridge Management Systems," *NCHRP Report 300,* Transportation Research Board, National Research Council, Washington, D.C., 1987.

[Hudson 87a]   W. R. Hudson, and W. Uddin, "Future Evaluation Technologies: Prospective and Opportunities," *Proceedings, 2nd North American Pavement Management Conference,* Toronto, Ont., 1987.

[Hudson 87b]   W. R. Hudson, C. Boyce, and N. Burns, "Improvements in On-System Bridge Project Prioritization," *Research Report 439-1,* Center for Transportation Research, The University of Texas at Austin, 1987.

[OECD 87]   *Pavement Management Systems,* Road Transport Research, Organization for Economic Cooperation and Development (OECD), Paris, France, 1987.

[O'Day 84]   D. K. O'Day, "Aging Water Supply Systems: Repair or Replace," *Infrastructure Maintenance and Repair of Public Works, Annals of the New York Academy of Sciences,* vol. 431, Dec. 1984, pp. 241–258.

[PIARC 95]   "Road Bridges," *Reference 20.11.B, Report of the Committee, 20th World Road Congress,* Permanent International Association of Road Congresses (PIARC), Montreal, P.Q., Sept. 1995.

[Pillared 96]   J. Pillared, C. Baumert, and M. Green, "ASCE Standards on Structural Condition Assessment and Rehabilitation of Buildings," *Standards for Preservation and Rehabilitation, ASTM STP 1258,* American Society for Testing and Materials, Philadelphia, Pa., 1996, pp. 126–136.

[Shahin 90]   M. Y. Shahin and J. A. Walter, "Pavement Maintenance Management for Roads and Streets Using the PAVER System," *USACERL Technical Report M-90/05,* Champaign, Ill., 1990.

[Uddin 86]   W. Uddin, A. H. Meyer, and W. R. Hudson, "Rigid Bottom Considerations for Nondestructive Evaluation of Pavements," in *Transportation Research Record 1070,* Transportation Research Board, National Research Council, Washington, D.C., 1986, pp. 21–29.

[Uddin 96a]   W. Uddin, R. M. Hackett, P. Noppakunwijai, and Z. Pan, "Three-Dimensional Finite-Element Simulation of FWD Loading on Pavement Systems," *Proceedings, 24th International Air Transportation Conference,* Louisville, Ky., 1996, pp. 284–294.

[Urban 84c]   H. P. Harty and B. G. Steinthal, *Guides to Managing Urban Capital Series, vol. 4: Guide to Selecting Maintenance Strategies for Capital Facilities,* The Urban Institute, Washington, D.C., 1984.

[Uzarski 88]   D. R. Uzarski and D. E. Plotkin, "Interim Method of Maintenance Management for U.S. Army Railroad Track Network," in *Transportation Research Record 1177,* Transportation Research Board, National Research Council, Washington, D.C., 1988, pp. 84–94.

Performance Modeling and Failure Analysis

8.1 Performance Evaluation

8.1.1 Performance concepts

An infrastructure facility is intended to serve its owners and users to certain pre-established levels of quality. The constructed level of quality usually does not stay constant over time and deteriorates throughout the service life. Based on the discussions of service life and performance in Chapter 3, a general definition of *performance* related to an infrastructure facility would be "the degree to which a facility serves its users and fulfills the purpose for which it was built or acquired, as measured by the accumulated quality and length of service that it provides to its users." Figure 8.1 illustrates the concept of *performance* where performance is represented by plotting a time series of a quality measurement, such as condition index (CI), on a scale of 100 (as for a new facility in the best possible condition) to 0 (totally unacceptable or failed). Of course other scales may also be used, depending on the facility, the measure, and other factors. A common scale is the 0–5 scale of the Present Serviceability Index for pavements.

In-service evaluation and condition assessment of infrastructure facilities, shown in Figure 8.1, have been discussed in detail in Chapters 6 and 7. The area under each curve in Figure 8.1 represents the accumulated service or performance. Using this graphical interpretation, several concepts associated with performance are identified:

- Initial condition, level of service, and quality are near the top of the scale in Figure 8.1 (i.e., 100), similar to a newly constructed facility after successful acceptance testing and commissioning. A minimum

Figure 8.1   Conceptual illustration of several forms of *performance* curves.

acceptance level must also be set. In Figure 8.1 this is illustrated as 40, but of course it will vary with many factors, such as class of facility, agency policy, safety, and economics.

- There is a deterioration of condition with time. This may reflect not only deterioration in physical condition, but also level of service, quality, function, structural capacity, and/or safety.

- The facility may or may not reach a minimum acceptable level before the end of the life cycle or analysis period. This can happen because of continuing deterioration, functional obsolescence, safety reasons, or a catastrophic event like fire, flood, earthquake, or any other disaster.

- The slope and form of the condition deterioration history determines the performance. A good performing facility should provide a high service level and should remain in relatively good and acceptable condition for most of the service life, such as shown by Curve A in Figure 8.1 (curvilinear form), Curve B (linear form), or Curve E (discontinuous form). Curve C shows a poorer performing facility due to a relatively higher deterioration rate in the initial years.

- The in-service evaluation and review of performance trends can lead to an effective intervention policy for maintenance, rehabilitation and renovation, and replacement or reconstruction (M,R&R) based on a preselected level of deterioration. A major restoration and rehabilitation intervention can extend the service life of a facility, as Curve D in Figure 8.1 does to Curve C.

A facility is the result of interrelated components, designed and built in a logical sequence, that deliver a final product that serves its intended purpose. Deterioration of physical condition is a complex process, as shown by wear and aging due to loading history and usage, degradation of materials of construction as affected by the environment, and the interaction of these mechanisms. Traditional design practices have relied on a preselected service life using safety factors to take into account uncertainties in design and demand inputs, without the concept of performance, as shown by the curves in Figure 8.1. This old practice is unrealistic, and appropriate performance measures should be identified and monitored so that historical data can be used to develop and improve performance-prediction models.

8.1.2 Deterioration process

A number of factors involved in the physical deterioration of a facility are listed in Table 6.1 in five major categories: load, environment, material degradation, construction quality, and interaction affects and other mechanisms. The table also lists primary and secondary condition deterioration mechanisms associated with possible surface defects, deformation, cracking or breaks, and failures. Deterioration is a function of properties of the materials used in the facility, demand on the facility, and the operating environment. The properties of a material can be grouped into physical, chemical, mechanical, thermal, and hydraulic. The behavior of the material can be further characterized under static and dynamic loading conditions. Chemical actions and additives associated with a given environment can produce drastically different changes in these properties. For example, addition of lime and fly ash to cement in ordinary Portland cement concrete can increase the compressive strength markedly. Polymer modifiers in asphalt can significantly decrease creep strain and therefore improve rutting resistance of asphalt pavements. Special alloys can reduce corrosion of steel. On the other hand, aggregates showing alkali-silica reaction will have durability problems. Saline environments and the use of salt on bridge decks lead to steel corrosion in steel and reinforced concrete structures. Wood structures will have problems related to water and termites. It is possible to develop reliable material models and predict the performance of a material under specific operating conditions and environment. An empirical model can be developed based on the accumulation of experience with that material, such as wood, steel, and concrete in different environments.

To develop a performance model, it is imperative to identify the deterioration mechanisms and deterioration time line. For many situations a curvilinear, S-shaped curve, as shown by Curve S in Figure 8.2, can be used to define performance. Various parts of the S-shape character-

Figure 8.2   Effect of M,R&R action on performance.

ize the deterioration process. Based on the systems concepts for pavement management [Haas 94] and for building assessment [Harris 96], the deterioration process and performance curve can be characterized by three distinct phases of the facility life in the absence of any major M,R&R treatment. The phases I, II, and III are shown in Figure 8.2 for a typical performance curve, Curve S. It has a small slope in Phase I, followed by accelerated deterioration in Phase II, and then a decelerated deterioration rate in Phase III (which is often the result of increased maintenance). Throughout the life the slope is never zero or negative, and the upper and lower bounds are known.

Phase I of the performance curve commences immediately following construction, or when all the components are new, or after a major M,R&R as in the case of Curve 2. The slope of the curve and time period in this phase will influence the overall performance. A smaller slope and longer period indicates good performance. The rate of deterioration and duration of Phase I is dependent upon the inherent properties of the construction materials used, the quality of construction, demand on the facility, and the interaction with the operating environment. The operators of the facility would like to see this period equal to the expected design life, ideally with minimum maintenance. A higher rate of deterioration will indicate poor performance and the possibility of early retirement before reaching the design life. A sharp increase in the rate of deterioration indicates the start of Phase II and possible deterioration of one or more structural components. Without appropriate maintenance and repair, the deterioration will accelerate in a relatively shorter period to the minimum acceptable level. This initiates Phase III—decelerating deterioration with incipient functional and structural failure. The condition is generally stable because it cannot get much worse. Some examples of this last phase are a badly cracked and potholed road, a silted dam, a nearly unusable building, and an unsafe tunnel. An earlier intervention, at the beginning of Phase III as shown in Figure 8.2, usually would be desirable. If deterioration is allowed to continue to the end of Phase III, reconstruction or replacement may be the only effective option.

8.1.3 Need for performance evaluation

Performance evaluation of a facility in an infrastructure management system is essential to

- Assess the service quality and functionality of the facility related to the materials and construction methods used
- Predict the rate of deterioration and possible failure of the facility
- Verify performance predictions and improve prediction models
- Measure the effectiveness of M,R&R strategies and value of extended life
- Update and improve life-cycle cost and benefit analysis methodologies
- Schedule M,R&R strategies and needs for future in-service inspections and evaluations
- Plan M,R&R budgets for short-term and long-term work programs and update network improvement programs

The development and study of performance models have received major attention from public works agencies, which are responsible for preserving the transportation infrastructure and providing expected service to the traveling public at local, state, regional, and national levels. In particular, a number of pavement performance models have been developed over the last three decades [Haas 94], and a national pavement performance database has been developed through the Strategic Highway Research Program (SHRP) conducted from 1987 to 1993 [SHRP 92, Uddin 95].

8.1.4 Service life

Prediction of service life is an important and fundamental aspect of performance modeling. This may or may not coincide with the design life. For an infrastructure facility constructed with different materials for different structural components and subjected to varying demand

and environment conditions, the service-life prediction can be complex. Therefore, a facility may have to be partitioned into several parts or segments based on their function, material, construction, and operating conditions. A logical partitioning of the facility may lead to predictable estimates of expected service life in terms of functional life. Traffic marking and sign posts on a road, painting of steel girders and other components of steel in bridges, and painting of the exterior of residential buildings are examples of the use of functional life span that can be established using an empirical database. It may not be necessary to establish a performance equation for these examples. For many widely used materials and construction methods, the expected service-life estimate is available from prior experience. However, the overall service life of a facility is a complex matter. Service life is the period of time from the completion of the facility to when the facility or any of its components reach a state where the facility cannot provide acceptable service because of physical deterioration, poor performance, functional obsolescence, or unacceptably high operating costs. Some typical expectations of infrastructure service life based on Table 2.1 are listed in Table 8.1. These lives are mostly 100 years or less, with the exception of landfills that are legally mandated to be safe for 250 years [Krinitzsky 93]. Table 3.2 shows a comparison of service lives of buildings based on codes and recommendations extracted from relevant documents from Japan, Canada, and the U.K.

Evaluation of overall service life of infrastructure assets should be based on the life of critical structural components. Service life can be estimated from: (1) empirical experience; (2) historical database using survivor techniques; (3) established performance models; (4) laboratory testing; and (5) accelerated field testing. The prevailing load, material, and method of construction, as well as environmental, ground,

TABLE 8.1   Typical Expectations of Infrastructure Service Life

Infrastructure facility and components		Expected service life
Airports	Buildings/structures	50–70 years
	Runways/taxiways/aprons	40–50 years
Bridges	Decks	20–30 years
	Substructure/superstructure	50–100 years
Dams, tunnels	For auto traffic, water	80–100 years
Nuclear power plants	—	20–40 years
Ports, marine terminals	—	70–100 years
Public buildings	—	30–100 years
Electricity transmission/ telephone lines	—	50–100 years

and operating conditions should always be considered to estimate the service life.

8.2   Performance Modeling

Performance modeling is an important part of infrastructure management at both project and network levels. A performance model relates a selected performance indicator to a set of causal variables such as age, load, load repetitions, usage history, material properties, environmental factors, and M,R&R history.

8.2.1   Performance indicators

Overall functional performance of a facility is generally assessed from the user perspective, with such measures or indicators as quality or level of service, system effectiveness, productivity and efficiency, and resource utilization and cost-effectiveness [Urban 84b]. On the other hand, the technical evaluation of performance is vital to engineers because it includes measures of mechanistic behavior and physical deterioration that lead to the correct selection of M,R&R alternatives. Therefore, it is important to include both user and engineer evaluation to formulate a comprehensive performance indicator. Humplick and Paterson recommend that the performance indicators for road-transportation infrastructure should include measures of sufficiency, efficiency, and effectiveness, as well as institutional aspects of effectiveness [Humplick 94]. Performance indicators can be grouped into the following broad categories:

- Service and user rating
- Safety and sufficiency
- Physical condition
- Structural integrity/load-carrying capacity

To establish an overall performance indicator for an infrastructure facility, the highest weight should be given to structural integrity and service to users. Some examples of performance indicators are provided in Chapter 3. Table 8.2 shows specific examples of performance indicators as related to the above categories.

8.2.2   Modeling techniques

Performance models can be developed by a variety of techniques, including the following:

TABLE 8.2   Examples of Performance Indicators

	Performance Indicators			
Facility	Service and user rating	Safety and sufficiency	Physical condition	Structural integrity/ load capacity
(1) Highways, roads, streets, parking areas	Present Serviceability Rating (PSR), ride quality, vehicle operating costs	Ratings based on skid resistance, accidents, congestion, and pollution	Present Serviceability Index (PSI), International Roughness Index (IRI), Pavement Condition Index (PCI) based on distresses	Structural rating based on deflection testing, remaining life, load capacity
(2) Airports: pavements, buildings, aviation facilities	Ride quality, user satisfaction ratings for other facilities	Operating ratings for aircraft, passengers and baggage, flow and capacity	PCI (pavements), Ride Quality Index (RQI); see (8) for buildings, air-traffic-control facilities, airport lighting, etc.	Structural rating based on deflection testing, remaining life, load capacity; see 8 for buildings and other facilities
(3) Bridges: interchange, overpass, underpass, rail/river crossing	Inventory and appraisal ratings for deck and other components	Ratings for safety, capacity, sufficiency, and congestion	Ratings based on distress and appraisal data, overall condition rating	Ratings of structural components based on visual inspection, vibration and seismic testing, nondestructive evaluation, load rating, and remaining life
(4) Railroad: terminals, track, bridge	Track quality rating, user satisfaction ratings	Ratings for safety, crossing accidents, operations, and capacity	Track Quality Index (TQI), Track Structure Condition Index (TSCI); see (3) for bridges and (8) for buildings	Track modulus, load rating, remaining life; see (3) for bridges and (8) for buildings
(5) Water: treatment plants, storage, supply	Water quality, user satisfaction ratings	Ratings for health hazard, water pressure, supply, consumption, and operations	Number of breaks per year, leakage per year, and number of repairs per year	Failures, damage to roads and other structures, remaining life, and demand

TABLE 8.2 Examples of Performance Indicators (*Continued*)

Facility	Performance Indicators			
	Service and user rating	Safety and sufficiency	Physical condition	Structural integrity/ load capacity
(6) Waste water and sewer	Treatment quality, user satisfaction ratings	Ratings for health hazard, discharge, capacity, and operations	Number of breaks per year, leakage per year, and number of repairs	Failures, damage to roads and other structures, remaining life, and demand
(7) Solid-waste facilities	Odor and treatment, user satisfaction ratings	Ratings for health hazard, fire hazard, discharge, capacity, and operations	Ratings for spilled trash, missed collections, soil and water contaminations, and facility condition	Failures, damaged and inadequate collection and disposal facilities, remaining space, and demand
(8) Buildings: Multipurpose complexes	Discomfort and user satisfaction ratings	Ratings for fire hazard, utilities, occupancy, capacity, and operations	Interruptions in utility services, heating and cooling effectiveness, air-quality rating, and condition ratings	Structural and ground failures, posting, remaining life, safety hazards, and demand
(9) Power-supply facilities	Ratings for interruptions in service, user satisfaction	Ratings for fire hazard, operations, and power generated and distributed	Frequency of service breaks, power surges, condition ratings of fixed assets; see (1) for roads and (8) for buildings	Failures, remaining life, safety, demand; see (1) for roads and (8) for buildings and other facilities

- Expert system incorporating a knowledge base of empirical experience
- Regression analysis
- Markov transition probabilities
- Artificial neural networks analysis
- Bayesian methodology
- Econometrics methods

Expert systems and regression-analysis techniques were discussed in Chapter 4, Sec. 4.4. A more comprehensive discussion of knowledge-based expert systems technology (KBES) applications, comprising an entire chapter, has been provided by Haas et al. in *Modern*

Pavement Management [Haas 94]. The authors illustrate that a significant feature of KBES, when compared to conventional techniques, is its ability to use knowledge when it is needed.

A large body of information is available on regression modeling techniques, ranging from relatively simple and introductory texts to advanced books. The reader can access any technical library to see the numerous texts that are available. Generally, regression analysis for performance modeling utilizes historical data to develop a relationship between an independent variable, such as condition index, and a set of independent or causal variables, such as those noted in Sec. 8.1.2. For network-level purposes, simplified time-series models may be sufficient, whereas on a project-level basis a more comprehensive set of data on materials properties, loads, environment factors, and so forth, may be needed to develop a sufficiently accurate model.

Markov transition probability models are particularly useful where a historical database does not exist. They capture the experience of engineers or technologists in a structured way by utilizing different classes or combinations of situations and condition states. For example, one class for developing a road-deterioration prediction model might be high traffic combined with a minimum thickness structure combined with a strong foundation. Of course, the combined boundaries for the variables comprising the classes have to be provided (say high traffic is greater than 10,000 AADT). Condition states, using the vertical scale example of Figures 8.1 and 8.2 and a CI measure could be $CI_1 = 100$ to 90, $CI_2 = 90$ to 80, and so forth (10 condition states). Many references are available on Markov modeling, including applications in the transportation field [Li 96].

Bayesian modeling techniques have existed for quite a number of years, and have been applied in a variety of infrastructure areas. Essentially, they use prior knowledge and/or experience (preposterior), which can be combined with actual observations (usually in cases in which such data are limited) to produce posterior estimates of deterioration [Larson 69]. Figure 8.3 illustrates the general procedure in the Bayesian approach. An example of extensive use of this approach is represented in the long-term pavement performance modeling in the SHRP studies [C-SHRP 95]. The main advantage of using a Bayesian approach, compared to classical regression analysis, is that a comprehensive database is not needed.

Artificial neural networks (ANN) analysis, which has gained considerable popularity in engineering modeling in recent years, involves parallel computational models for knowledge representation and information processing [Begley 95, Ghaboussi 92]. The processing units model the action of the synapses and dendrites in the human brain. Because of their fundamental similarity to the human brain in

Figure 8.3 General approach for Bayesian modeling.

terms of hardware, neural networks have some unique, human-like capabilities in information processing. Neural networks are capable of learning complex, highly nonlinear relationships and associations from a large body of data. The information and the knowledge learned by the neural networks is encoded and stored in the connection weights of the network. The retrieval of the stored information is done routinely by providing the network with an input pattern, which acts as a key. Detailed discussion on ANN modeling is provided in Chapter 4, Sec. 4.4.

Econometrics methods can also be effectively used in certain situations to predict performance [Madanat 95]. The authors contend that simple regression models, which are commonly used to model deterioration or distress propagation, suffer from selectivity bias due to the nonrandom nature of the estimation sample used. An example application has involved pavement-cracking progression, using a method known as Heckman's procedure and a full information maximum likelihood method, both to correct for selectivity bias.

8.2.3 M,R&R intervention performance modeling

During the life cycle of an infrastructure facility, an M,R&R intervention may be required, as illustrated in Figure 8.2. The following key alternative intervention actions are possible:

- Do nothing, or defer M,R&R action beyond the minimum acceptable level until the expected time of replacement
- Corrective maintenance to keep the condition at the minimum acceptable level

- Repair, rehabilitation, restoration, and renovation to improve the condition and extend the life at or before the minimum acceptable level is reached
- Replacement and reconstruction

It is generally assumed that routine and minor emergency maintenance does not have any significant effect on condition and only a major M,R&R action improves the condition, as shown by Curve 2 in Figure 8.2.

Given the foregoing intervention alternative actions, the question is how to model the postintervention performance. Basically, the same techniques identified in Sec. 8.2.2 are applicable. Quite often it is sufficient to simply use the same performance trend or model for Curve 2 as for Curve S in Figure 8.2, especially if the basic deterioration pattern of the facility is not expected to change. Of course, if a major structural or functional improvement is achieved by the intervention, then a new Curve 2 model would likely be needed, even though the same basic techniques still apply.

8.3   Failure Analysis

Failure analysis is an important way to examine a failed facility or its components and determine the possible reasons of an unexpected failure. Some failures are catastrophic, resulting in losses of human lives and property. Examples are bridge collapses, off-shore oil-rig disasters, foundation failures, failure of dams and dikes, corrosion-related failures, and earthquake damages. Sometimes a full-scale instrumented facility is intentionally failed, to learn the failure mechanisms and their interaction effects. Failure analysis is also the most important component of forensic engineering [Bodzay 91].

8.3.1   Approach and technologies
for failure analysis

Failure analysis requires a collaborative approach among structural and design engineers, material scientists, and metallurgists, as well as the use of the latest available nondestructive evaluation (NDE) methods. As discussed earlier, the normal deterioration process starts soon after construction, and deterioration is generally not visible for some period of time. A catastrophic failure is triggered by a sudden increase in critical stress, strain, displacement caused by corrosion, dynamic loads and shock waves, hydraulic pressure, and ground failure.

The knowledge of the precise reasons of failure can be used to improve performance models. Environmentally-associated cracking and stress-corrosion cracking (SCC) play important roles in lifetime predictions because their catastrophic character can affect a wide variety of metals and alloys exposed to specific environments [Roberge 95]. It is now believed that SCC, fatigue, and general fracture should not be considered separately, because there is a continuum in their relative failure boundaries as shown in Figure 8.4. An expert-systems approach may be used for failure prediction in this case because of the lack of any obvious relationship between these phenomena.

A thorough examination of the design criteria, construction details, service conditions, and usage history provides the clues to unexpected and premature failures. Therefore, the inventory and in-service monitoring and evaluation principles discussed in Chapters 4 through 7 are important for failure analysis. Inadequate analysis of soils and ground conditions is a primary reason for common slab cracking and settlement; for example, the slab-on-grade foundation problems in the regions of expansive clay soils. Many costly failures on expansive soils result from a failure to consider expansive soil movements or the failure of the builders to incorporate the design considerations for structural integrity of the foundation on expansive soils [Godwin 93]. As one moves to new materials frontiers and tries to avoid catastrophic failures, one needs to rely more on advanced NDE technologies, analysis methods, such as fracture mechanics and neural network

(log) flaw size

Figure 8.4   Relative failure boundaries among fracture, SCC, and fatigue. [Roberge 95].

analysis, and advanced computational tools like three-dimensional finite-element analysis. The following sections present some case studies of failure analysis.

8.3.2 Lessons from the Schoharie Creek bridge failure

The Schoharie Creek bridge built in the early 1950s on the New York State Thruway collapsed on April 5, 1987, in a few hours during floods [Thornton 88]. As the authors noted, Pier 3 failed by scour due to stream-bed erosion, dropping spans 3 and 4 into the creek followed by failures of other piers and spans. A thorough investigation of this failure indicated poor communication between the designer's original assumptions of soil properties and riprap and the construction and maintenance teams, as well as inadequate inspections. Some of the key recommendations from the study are [Thornton 88]:

- An expanded bridge-inspection program by engineers familiar with bridge design

- A structural integrity check on a periodic basis (at least once every ten years)

- Inspection on at least the following schedule: (1) complete structure and foundation once a year; (2) water crossings and stream-bed changes after each storm; (3) movements of bearings and expansion joints during at least one high-temperature period, low-temperature cracking of tension members or flanges, and drainage of salt-laden run-off during a low-temperature period (every year); and (4) after each accident on the bridge

- Thorough and supervised underwater inspection

- A scour indicator system for foundations with running water (checked during each inspection)

- Review of all maintenance reports and repair/rehabilitation records before and after their implementation

- Frequent contact with the bridge owner for knowledge of deficiencies and urgency of required repair

Scouring and erosion of stream beds around piers and columns are primary reasons for the undermining of bridge foundations and resulting bridge failures. The mechanisms and analysis of scouring and erosion of rocks and soils have not received adequate attention in the past. A new quantitative methodology for an erodibility index, proposed by Annandale et al., depends on the erodibility threshold of a material and the erosive power of water [Annandale 96].

8.3.3 Fire and explosion investigation

Calligiuri describes the following investigation of a fire and explosion case [Calligiuri 91]. The initial analysis of the remaining evidence of the investigation provided no immediate reason for the accident; however, attention was given to the crushed segment of a natural gas pipeline that ran beneath the explosion site. Detailed visual examinations, metallurgical tests, and pressure vessel tests were conducted on undamaged pieces. Welds were examined by ultrasonic testing and video probe. Computer simulations of the buried pipe were conducted by the DANTE finite-element code for no internal pressure and 2,067 MPa (300 psi) internal pressure. The computed deformations with time supported the conclusion that the pipeline was depressurized prior to the fire [Calligiuri 91].

8.3.4 Lessons from earthquakes

Most of the current earthquake-resistant design codes and retrofitting of older structures have been based on in-depth postearthquake inspections and structural seismic analysis of damaged facilities. However, newer earthquakes bring newer concerns and provide opportunities to assess the performance of existing procedures. In an award-winning article "The Lessons from Kobe" in the *Los Angeles Times,* Hotz and Reich summarize the results of their interviews with many earthquake experts after the 6.8 Richter scale earthquake on January 17, 1995 in Kobe [Hotz 95]. The magnitude of the 1994 Northridge earthquake in Los Angeles measured 6.7 on the Richter scale. Table 8.3 is based on the above article.

8.4 Summary

This chapter has only scratched the surface on performance modeling and failure analysis. Each infrastructure type requires its own data and modeling. Theoretical equations, such as stress and fatigue, can help in analysis of the structure, but always falls short in modeling the total life cycle. More detailed modeling information is available for all fields of infrastructure.

Failure analysis adds information to the experience base, particularly in defining end points and critical time elements. Detailed examination histories are available in each field.

References

[Annandale 96]   G. W. Annandale, S. P. Smith, R. Nairnes, and J. S. Jones, "Scour Power," *Civil Engineering,* July 1996, pp. 58–60.
[Begley 95]   E. F. Begley and C. P. Sturrock, "Matching Information Technologies with

TABLE 8.3   The Lessons from the Kobe Earthquake (after [Hotz 95])

Concerns	What Happened in Kobe	Japan's Preparedness	California's Preparedness
Highways	The ground shaking exceeded design limits by so much that hundreds of support pillars were seriously damaged. Four railways and three major highways were cut.	Huge brittle bridge and highway supports are designed to withstand shaking at about one-third the force of gravity. In the Kobe quake, the force exceeded that limit and pillars snapped.	Highway engineers emphasize smaller, more flexible supports that can sway with ground shaking up to 70% of the force of gravity (g) without breaking. In the Northridge earthquake, shaking exceeded 1 g in some areas.
Steel-frame buildings	Most newer high-rise buildings appeared to withstand the shocks, while precode buildings collapsed in many cases.	High-rise buildings constructed after 1981 incorporated flexible design meant to safely absorb seismic energy, but are still stiffer than in California.	Flexible-frame design is favored. After the Northridge earthquake, structural engineers found dangerous cracks in more than 120 buildings. The newest buildings with shock-absorbing foundations, called base-isolation systems, emerged without damage.
Water and gas	Miles of water, sewer, and and gas pipes broke. That, compounded with Kobe's high population density, resulted in 300 fires, twice as many as occurred after the Northridge earthquake.	Japanese utility companies had developed what they believed would be an automatic cutoff system to safely shut down natural gas lines in the event of a serious earthquake.	Unprotected oil and gas lines, along with sewer and water systems, remain a major concern.
Routine residential construction	Thousands of homes simply collapsed, killing or injuring the occupants.	Traditional Japanese housing relies on a frame of a few large posts topped by a heavy tile roof, with no cross-bracing.	In California, average housing uses cross-braced, stud-wall construction, with plywood shear panels lending lateral strength.
Person readiness	Few people seemed to have personal earthquake preparedness kits with emergency food, water, and medical supplies.	In Japan, there are annual earthquake drills, but fewer than 30% of the people in Tokyo participated in last year's exercises; fewer than 10% in Kobe participated.	For all practical purposes, formal emergency drills in businesses and schools are nonexistent. However, personal earthquake kits are increasingly common in California homes and workplaces.

the Objectives of Materials Data Users," *Computerization and Networking of Materials Databases*, Vol. 4: *ASTM STP 1257*, American Society for Testing and Materials, Philadelphia, Pa., 1995, pp. 253–280.

[Bodzay 91]   S. Bodzay, "Failure Analysis in the Forensic Domain," *Conference Proceedings of the International Conference and Exhibits on Failure Analysis*, Montreal, P.Q., July 1991, pp. 249–252.

[Calligiuri 91]   R. D. Calligiuri, "Failure Analysis, Prevention, and Testing," *Conference Proceedings of the International Conference and Exhibits on Failure Analysis*, Montreal, P.Q., July 1991, pp. 191–197.

[C-SHRP 95]   Canadian Strategic Highway Research Program, "Bayesian Modeling: Joint C-SHRP/ Agency Applications," *Technical Brief No. 8, C-SHRP*, Transportation Association of Canada, Ott., Nov. 1995.

[Ghaboussi 92]   J. Ghaboussi, "Potential Application of Neuro-Biological Computational Models in Geotechnical Engineering," *Numerical Models in Geomechanics*, Pande and Pietruszczak (eds.), Rotterdam, the Netherlands, 1992.

[Godwin 93]   C. J. Godwin, "Slab-on-Grade Foundations on Expansive Soils," *Building Research Journal*, National Consortium of Housing Research Centers, Vol. 2, No. 2, Nov. 1993, pp. 33–43.

[Haas 94]   R. Haas, W. R. Hudson, and J. P. Zaniewski, *Modern Pavement Management*, Krieger Publishing Company, Malabar, Fla., 1994.

[Harris 96]   S. Y. Harris, " A Systems Approach to Building Assessment," *Standards for Preservation and Rehabilitation, ASTM STP 1258*, American Society for Testing and Materials, Philadelphia, Pa., 1996, pp. 137–148.

[Hotz 95]   R. L. Hotz and K. Reich, "Lessons of Kobe: Hope, Caution," *Los Angeles Times*, Feb. 12, 1995.

[Humplick 94]   F. Humplick and W. D. O. Paterson, "Framework of Performance Indicators for Managing Road Infrastructure and Pavements," *Proceedings, 3rd International Conference on Managing Pavements*, National Research Council, Vol. 1, May 1994, pp. 123–133.

[Krinitzsky 93]   E. L. Krinitzsky, "The Hazard in Using Probabilistic Seismic Hazard Analysis," *Civil Engineering*, Nov. 1993, pp. 60–61.

[Larson 69]   H. J. Larson, *Introduction to Probability Theory and Statistical Inference*, 2nd ed., John Wiley & Sons, New York, 1969.

[Li 96]   N. Li, W. Xie, and R. Haas, "Reliability-Based Processing of Markov Chains for Modeling Pavement Network Deterioration," accepted for publication in *Transportation Research Record*, Transportation Research Board, National Research Council, Washington, D.C., 1996.

[Madanat 95]   S. Madanat, S. Bulusu, and A. Mahmoud, "Estimation of Infrastructure Distress Initiation and Progression Models," *Journal of Infrastructure Systems*, Vol. 1, No. 3, Sept. 1995, pp. 145–150.

[Roberge 95]   P. R. Roberge, "Background and Basis for a Knowledge Elicitation Shell for Lifetime Predictions from Stress Corrosion Cracking Data," *Computerization and Networking of Materials Databases* Vol. 4, *ASTM STP 1257*, American Society for Testing and Materials, Philadelphia, Pa., 1995, pp. 136–150.

[SHRP 92]   *SHRP Product Catalogue*, Strategic Highway Research Program, National Research Council, Washington, D.C., 1992.

[Thornton 88]   C. H. Thornton, R. L. Tomasetti, and L. M. Joseph, "Lessons from Schoharie Creek," *Civil Engineering*, May 1988, pp. 46–49.

[Uddin 95]   W. Uddin, "Pavement Material Property Databases for Pavement Management Applications," *Computerization and Networking of Materials Databases*, Vol. 4: *ASTM STP 1257*, American Society for Testing and Materials, Philadelphia, Pa., 1995, pp. 96–109.

[Urban 84b]   S. R. Goodwin and A. E. Peterson, *Guides to Managing Urban Capital Series*, Vol. 2: *Guide to Assessing Capital Stock Condition*, The Urban Institute, Washington, D.C., 1984.

Concepts of Total Quality Management

Design for Infrastructure Service Life

9.1 Introduction

In Chapters 1 and 2, the interrelationship of design with infrastructure management was defined. Design is a uniquely project-level activity of infrastructure management, along with construction, maintenance, and rehabilitation. Planning, budgeting, and programming of maintenance and rehabilitation can be done at the network level, but design is applicable to a specific unit of infrastructure or project.

The process flows from design to construction, maintenance, and rehabilitation. For the infrastructure to function effectively, it must start with good design. In many types of infrastructure, such as pavements, bridges, and buildings, there is talk of theoretical design versus empirical design. This implies, on the one hand, that the choice of physical elements is totally based on theory, or usable mechanistic theory; on the other hand, that the selection or sizing of elements is based entirely upon empiricism or experience. In truth, all design is somewhat subjective, since factors such as criteria, design life, and level of service, all change from location to location and from agency to agency.

This chapter is not a "design guide." It is impossible to provide such a design guide in one book, since each type of infrastructure has its own design requirements, and within a given infrastructure type each agency has its own specific design elements and approach. Rather, this chapter covers the design process, and the technology and concepts of design. Certain aspects of design are common to any facility and any type of infrastructure. Concepts of design range from visual, such as the design of an automobile or architecture design, to physical and mechanical, such as the design of an engine, an airplane, or a bridge.

Engineering of physical infrastructure is not as freeform as architectural design, but it does offer flexibility and an opportunity for creativity.

9.1.1   Reliability and design

Design is often practiced in a deterministic way, in which a specific design can be defined that uniquely fits a set of physical circumstances. Such an approach is inadequate because of the true variability that faces all designers in terms of inputs and service life of the facility. Bridges, for example, are designed on the basis of predicted traffic and vehicle loads, yet no one has good control over the actual service loads that will be carried on the bridge. Legal load limits vary with respect to bridge configuration, and vehicle operators do not always adhere even to legal limits. Furthermore, load zoning and characteristics are political decisions that can change during the life of a facility.

These uncertainties have generally been dealt with by inserting safety factors into the design process. In structural design, for example, the yield strength of the material is determined by test, but a working strength for design is obtained by dividing the yield strength by a factor of safety. For example, if the factor of safety is 2.0, and the yield strength is 276 MPa (40,000 psi), then the working strength would be 138 MPa (20,000 psi).

Such working-stress and factor-of-safety methods are outmoded for the modern design of infrastructure. Reliability concepts are far superior, and any infrastructure engineer is urged to examine reliability-based methods in detail for design. For example, in 1985 AASHTO went to a reliability-based method for pavement design, which has been successfully used [AASHTO 85].

9.1.2   Design technology framework

A general framework can be presented for design that is quite comprehensive. Most infrastructure design methods do not include all aspects of the framework; however, logical extension of past practices will lead to continuous improvement in the design for individual infrastructure types over time. Historically, design has often been compartmentalized; for example, water-pipe design or sewer design is often considered as the process of sizing a pipe or flow structure, when in reality it also should involve selection of material, optimum location of the physical unit, and other details aside from pipe size alone. This problem has certainly applied to bridges and pavements, where design has often been relegated to selection of layer thickness for pavements, or selection of structural-member size and weight for bridges. In bridges, for example, custom within an agency may dictate the use of pre-stressed concrete members so that the designer overlooks the possibility of comparing other material types, such as steel, or other forms, such as arches. In pavements, it is not uncommon for the agency to preselect a flexible or rigid pavement, so that the so-called design process merely becomes a function of selecting thickness. Modern technology and computer tools make it completely possible for the designer to consider a fully functional design process, and design education should be broadened in this regard. Since design is the first major cost-related element in infrastructure management at the project level, good economical optimum design is critical to any effective infrastructure management. Although it is possible to optimize the balance of infrastructure management given a suboptimal design, such a process will never be truly optimal if the design is shortchanged.

Figure 9.1 presents a comprehensive diagram of most design practices. It is generic, and does not cover the details of any specific method. Common to all methods, however, are inputs of loading (either traffic or use), existing environment, and available materials. These factors are handled in two ways. The most common historical method has been to take the input and compare the design for a set of common alternate designs, check these with design models to see if they satisfy requirements, and alter factors and try again until they do. The other option—or more modern method—is to enter the loading, environment, and available materials factors into a comprehensive design model to predict a reliable design for the required service.

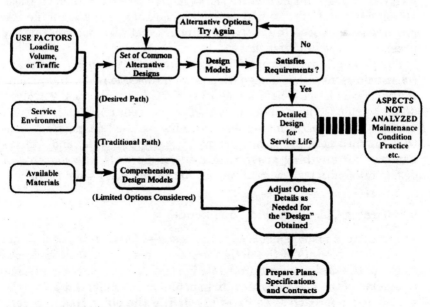

Figure 9.1   Description of most design practices.

In each case, standard practice has often been to adjust other details, such as connections, slab joints, joint spacing, and so forth, for the design obtained. This process then culminates with the preparation of plans, specifications, and estimates—the so-called contract documents. Not considered in this separate process are the aspects of maintenance, construction practices, and economics.

9.1.3 Evolution of design technology

The design process and methods have evolved greatly over the years. Most early designs were based on experience. These included the use of timber beams developed from local trees near the bridge site or building being constructed, the use of stone pilasters for construction of churches and large historical buildings, the intuitive use of large stone bases in the Roman road system, and the construction of stone aqueducts in ancient Greece and Rome.

However, integrated design is not a totally new development. A recent visit to ancient cities in Greece showed the integration of sewers and storm sewers into the paved block streets constructed for the main thoroughfare of the old (700 B.C.) city. Fastener holes were designed into some of the paving blocks to provide clean-out access to the sewers underneath.

It can be said that the quality of design, historically, often depends upon the intuition, knowledge, and hard work of the designer. The purpose of good infrastructure management is not to replace these strong characteristics, but to provide tools that enhance the design process so that less innovative engineers can also become quality designers in the broadest sense.

The general approach used by some agencies for many years has consisted essentially of designating "standard sections" for the various ranges of conditions. These standard sections have historically evolved differently for various environments. The inherent implication in such methods is that performance or service life will be "satisfactory" with the standard design, "unsatisfactory" for anything less, and "uneconomical" for anything greater. In modern methods, the concepts of design reliability have a great impact on this implicit assumption.

9.2   Design Objectives and Constraints

Infrastructure management helps the designer focus on the basic function and constraints of the design process. This avoids the rote application of the design procedure while ignoring other available alternatives. One of the first activities that should be accomplished under the systematic approach is to carefully define the objectives and constraints of the problem. Objectives and constraints apply to the facility and the design process, or to the designer's activities. Some apply to both.

9.2.1 Facility objectives

The principal objectives to be fulfilled by a facility, both during construction and in service, are economic and social in nature. They include but are not limited to the following:

1. Maximum or reasonable economy, in terms of both agency and user costs

2. Maximum or adequate safety

3. Maximum or reasonable serviceability over the design period (performance)

4. Maximum or adequate capacity for both the magnitude and number of repetitions of demand (loads, volumes, and flaws)

5. Minimum or limited physical deterioration due to environmental and use influences

6. Minimum or limited noise and air pollution and environmental disturbances during construction

7. Minimum or limited disruption of adjoining land or facility use

8. Maximum or good aesthetics

These objectives pose several major contradictions, which occur with any complex system or facility that attempts to fulfill social and economic needs. Consequently, each unit constructed represents some compromise or trade-off among objectives. The relative influence of the objectives in a particular design situation varies with such factors as rural or urban area, use, and environment.

The first three objectives listed are very important. Economy is a prime requisite, because most public agencies feel that they have insufficient funds available for all the desired investments. Safety is another major requirement the facility must fulfill. Serviceability should also be sufficiently high over the design period to ensure the desired level of quality to the user in terms of speed, operating costs, delays, and comfort.

The fourth objective, maximizing capacity, is applicable to both the facility and to the design process. This objective competes directly with the objective of maximizing economy. The fifth objective, minimizing physical deterioration, is related to serviceability in that deterioration is a major factor in causing serviceability loss. However, it is identified as a separate objective because it also competes with the economic objective in terms of maintenance and rehabilitation costs.

The last three objectives are usually of lesser importance, except in buildings. Aesthetics, or appearance, can be a significant objective for certain special applications, or in those cases where patching and repairs are extensive and appear ugly to users or adjacent property owners and observers.

9.2.2 Facility design objectives

In one sense, the objective of the design process is to design a facility that, when constructed, will accomplish the functional objectives previously defined. However, the specific objectives of the design process can be stated in technical, economic, and social terms:

1. Develop a design strategy of minimum (or reasonable) economy, safety, and serviceability.
2. Consider all possible design strategy alternatives.
3. Recognize and incorporate the variability in the design factors.
4. Maximize the accuracy of predictions of serviceability, safety, and physical deterioration for the alternatives considered.
5. Maximize the accuracy of estimating cost and benefits.
6. Minimize the costs of design, including labor, testing, and computer time.
7. Maximize information transfer and exchange between construction and maintenance personnel.
8. Maximize use of local materials and labor in the design strategies considered.

Some of the foregoing objectives correspond with the objectives for the facility itself, whereas others are peculiar to the design process. Most of the objectives compete with other objectives and require trade-offs.

The first objective simply reflects the need for the design to recognize the objectives of economy, serviceability, and safety. To accomplish this it is necessary, as stated in the second objective, to consider all the possible or feasible design-strategy alternatives. Because these can amount to several hundred combinations for any given situation, the use of computer analysis is necessary for accomplishing this objective.

The third objective, recognition of the stochastic nature of design, construction, and maintenance factors, should be important criteria for any design method. Early contributions to the inclusion of variability in the pavement design process were presented in the early 1970s [Darter 71, Darter 72, Darter 73a, Darter 73b]. There are significant parallel contributions in bridge and foundation design.

The fourth objective, maximizing the accuracy of the design predictions, is related to the third objective, and also to the quality of the input data. Decreased uncertainty in design predictions is important to planning activities so that investment programming can be accomplished more reliably.

The accuracy of the cost and benefit prediction, the fifth objective, is directly related to the accuracy of the design predictions, as well as the ability to estimate the construction, maintenance, and user costs.

Minimizing the costs of the design process in terms of materials testing, personnel time, and preparation of drawings, the sixth objective, is important, but it obviously competes with the other objectives. If the cost of the design process is underfunded, the result can be an incomplete, inadequate, or excessively costly design strategy.

The seventh objective, information exchange between those responsible for construction and maintenance, has unfortunately been neglected to a large extent in design practice. There have been instances in North America of maintenance crews unfamiliar with continuously reinforced concrete pavements using asphalt to fill the numerous hairline cracks that are a design feature of this pavement type and do not require maintenance. The resulting mess might be more readily ascribed to the designers' failure to communicate with the maintenance people than to the maintenance people themselves. It is extremely important that such communication occur, because proper construction and maintenance activities "carry" the design through the life of the facility.

The last objective, involving the use of local materials and labor, is perhaps a self-evident desirability for most situations. This objective may complement the first objective of maximum economy, but in some cases local materials may require stabilization or treatment.

9.2.3 Design constraints

The major design constraints are economic and physical or technical in nature and include the following:

1. Availability of time and funds for conducting the design and the construction
2. Minimum level of serviceability allowed before rehabilitation
3. Availability of materials
4. Minimum and/or maximum dimensions allowed
5. Minimum time between successive rehabilitations
6. Capabilities of construction and maintenance personnel and equipment
7. Testing capabilities

8. Capabilities of the structural and economic models available

9. Quality and extent of the design information available

The first constraint might involve not only a limit on the funds that can be applied to a particular project, but also the opportunity costs of those funds. All the other constraints, which are essentially physical or technological, are relatively self-explanatory. Each has economic implications, and is therefore related to the first constraint.

9.3   Design Framework and Components

In an infrastructure management system, the design phase involves several activities. These may be broadly classified as: (1) information needs related to inputs, objectives and constraints; (2) generation of alternative design strategies; and (3) analysis of structure, flow characteristics, and other aspects of the facility, along with economic evaluation and optimization of these strategies.

9.3.1   Information needs

The information needs of good infrastructure management are extensive. Detractors may cite this as a reason for not wanting to implement an infrastructure management system (IMS), which is a specious argument. The information needs are inherent in the problem, not in the IMS. They only are stated explicitly by the management process. The needs were always inherently there, and any failure to acknowledge and fill those needs means that a suboptimal design will be selected in most cases, or that the life of the infrastructure will suffer.

Much of the needed information can be obtained from network-level management activities, which generally take place prior to project-level design. It is important for the designer to make full use of available information from network-level activities and from associated databases. Failure to do so will result in wasted resources and suboptimal design at best. At worst, it will result in inadequate design and early failure. The designer must not, however, assume that network-level data quality is adequate for design. It merely serves as a starting point. It is essential to supplement network information with data collected specifically for the project being undertaken. Three main categories of data stand out: (1) environment; (2) loading and traffic; and (3) materials characteristics.

9.3.1.1   Environment.
Environmental data may be generalized, but a facility often exists in a unique location that requires characterization of the environment with more detail than is available on weather

maps, weather databases, or from other general sources. Depending upon the type of environment and the agency involved, experience should govern the way these environmental data are to be handled.

9.3.1.2 Projected use, loading, and traffic. Information on projected use, whether it be volumes of sewage flow, water consumption, average daily vehicular traffic or vehicular axle loads, is a matter of estimating and projecting future needs, rather than a matter of measuring current characteristics. This makes the process more difficult, and is an area where problems have often existed in past design. Historical records on similar facilities should be used, but consideration must also be given to potential changes that may occur as the situation evolves. Additionally, such projections should factor in demographic trends, population growth, economic development, land use, and other factors. These generally will be handled in a planning phase at the network level, but sharpening the estimates is required at the design phase.

9.3.1.3 Materials characteristics. It is impossible to design a facility effectively unless the service characteristics of the materials or elements to be used in the facility are understood. In many cases, the characteristics of the materials will change with environment and use, and in other cases the characteristics change with construction practice, availability of materials sources and other factors beyond the control of the designer. In some cases, the designers have assumed materials characteristics and left it to the construction phase to "provide the materials as designed." This is not a wise practice—materials must be considered effectively in the design process, even though the final function of providing the required materials must be accomplished during construction. Some materials degrade with exposure and/or with time and use due to fatigue. It is essential for the designer to consider any such material degradation.

Ultimately, material properties can only be obtained through appropriate physical testing. Such testing can be expensive and is too often ignored, but all designs depend on adequate knowledge of strength, stiffness, and deformation characteristics.

9.3.2 Other information factors required

Other factors to be considered in design include: costs, estimated service life, and various details required as input to a specific design method. Items such as inflation, user costs, and related topics may be difficult to obtain; however, this difficulty is not a valid reason for failure to make a reasonable estimate. Some designers, for example, ignore user costs and economic analysis because they say it is impossible to predict inflation and the exact costs to be borne by the user. This is a fallacy and can result in gross error. It is far better, for example, to assume a reasonable interest rate and inflation rate for a combined cost of money of 5 percent or 6 percent than it is to ignore the cost altogether. By ignoring the factor one is assuming the rate to be zero, which is obviously erroneous. Such factors can be handled in design through the use of sensitivity analysis for various ranges of user cost, inflation, and interest rates. This approach at least gives the designer a reasonable basis for making a final subjective judgment, or for providing the information to the decision-maker, which may be the administrator or a governing board, such as the city council.

9.3.3 Generating alternative design strategies

Designs of infrastructure involve more than the selection of a thickness or a beam dimension; therefore they are termed *strategies* and consist of a combination of elements, such as various configurations and future rehabilitation actions. They also include various material types and sources, expected construction, maintenance, and performance evaluation policies, as well as construction quality-control and assurance methodologies.

The need for a design alternative to specify materials type and dimensions is apparent. However, without including the expected construction and maintenance rehabilitation policies as part of the strategy, there may be appreciable error in the predicted performance outputs for that facility structure. Rehabilitation alternatives also become part of any design strategy if the service life of the facility drops to the minimum acceptable or terminal level before the end of the design period. An exception would be a maintenance policy that kept the facility at or above the minimum acceptable level of service to the end of the design period. Alternatively, if financial constraints prevail, maintenance at this level might continue only until funds were available for rehabilitation.

In formulating rehabilitation alternatives there are two major, interrelated aspects that must be considered:

1. Physical or structural aspects with respect to dealing with damage, lack of adequate service, lack of adequate safety, and so forth.

2. Policy aspects with respect to procedures for handling use or traffic during the rehabilitation or remodeling process. In the case of a building, Should the entire building be closed, or should the renovation be handled floor by floor? This decision will depend on how safety aspects are affected, and the type of renovation being undertaken. In a case such as the Pentagon, the headquarters of the U.S.

Department of Defense, the facility is so large that it would be physically impossible to close the entire structure for renovation at one time.

In summary, the design phase of infrastructure management will only be adequate if it considers a reasonable number of alternatives; thus, the process of generating alternatives is critical to the process of optimum design.

9.3.4 Analysis, economic evaluation, and optimization

Most design methods do not include all the activities of economic evaluation and optimization; rather, they include only analysis. However, design methodology has for the past 15 years been moving toward this more idealized form.

The first step in the analysis of any facility alternative is the application of appropriate structural and/or use models. In the case of hydraulic facilities, the models involve flow characteristics and other use factors. In buildings, bridges, and pavements, the primary models are often structural or functional.

In each case, the models must be sufficiently comprehensive to cover the needed areas of concern. The use of incomplete models can have enormous costs. As an example, there was a trend in the 1960s to use thin, lightweight concrete bridge decks to reduce the dead-weight of bridge structures, and thus reduce the size of major beam components. The resulting structures, however, deflected more under dynamic loads, and the thinner bridge decks were unable to sustain the additional deflection. As a result the U.S. interstate highway bridges failed earlier than expected, and underwent a massive replacement of decks in the 1980s.

A similar example occurred in continuously reinforced concrete pavements (CRCP). Application of structural models showed that a thin slab was sufficient to carry the load; however, no deflection analysis was run in comparison to the underlying subgrade. The large deflection of the thinner slab resulted in permanent deformation of the subgrade in many cases, which then caused loss of support and subsequent failure of the slab. This problem—and the failure to fully understand it—caused the cessation of the construction of CRCP in most of the world, even though it remains a very competitive construction practice in Texas and elsewhere where the deflection criterion has been added.

Current technology cannot adequately predict the type and degree of all distress modes that a facility will be subjected to as a function of time and use. Consequently, composite structural models are often used in an attempt to make a direct prediction of service quality versus age. Limiting values are often placed on other elements. The reader is encouraged to apply these general concepts and principals in reviewing the detailed design in his or her particular infrastructure type. The economic evaluation of a facility alternative should involve the assignment of costs and benefits to the predicted outputs in terms of the user, plus the cost of providing that alternative to the agency in terms of materials construction, maintenance, and rehabilitation. These variables are then incorporated into an economic model to determine the total costs and benefits of the strategy. (See Chapter 14 for a detailed presentation on life-cycle cost and benefit analysis.)

It is desirable to include benefits in the economic evaluation, but this is not currently the case in much of facility-design practice. One reason is that it is difficult, for example, to separate benefits for a pavement or bridge from overall highway benefits. Another is that the difference in user costs between alternative designs can also be considered as benefits but are difficult to determine. Research by the World Bank and subsequently by the U.S. Federal Highway Administration has successfully related vehicle operating costs to pavement conditions; other work is applicable to bridges and other types of infrastructure [Zaniewski 82, Chesher 85]. It is strongly urged that a complete economic analysis be made for all design alternatives, treated in detail in Chapter 14.

When all of the alternative design strategies have been analyzed and evaluated, optimization should be used to define the best strategy or strategies for presentation to the decision-maker. Frequently this is accomplished by comparing the discounted total costs for the alternatives, or, when comparing alternatives of different life, an average annual cost factor can be used to recommend strategy selection.

9.4 Design Effectiveness

Design usually involves a substantial degree of subjectivity, limited by rules of mechanics and physics, but oriented toward optimizing certain features as determined by a client and reflected in the stated criteria of the client. Many additional factors, such as schedule requirements, vendor capabilities, environment, and cost also must be incorporated into the design process. Nonetheless, design is perhaps the most central point of definition for a project in that ideas and information are transcribed to paper in the form of specific and coordinated instructions for the project's construction and documentation. Design effectiveness is related to studies of outputs, or products, of the design process, and is specifically limited to presenting a method for evaluating those outputs.

9.4.1 Measurement of design effectiveness

In recent years, increasing attention has been given to productivity and productivity improvement in the industrialized nations. In the United States, particular interest has focused on productivity in the construction industry, which is often criticized for slow technological growth and escalating costs.

Measuring design productivity is more difficult than measuring construction productivity. More and more people, however, are realizing that the real measures of the effectiveness of the design effort are found in the construction, start-up, and operational phases of a project. Thus, it may be more beneficial to develop a method for evaluating the effectiveness of a design, rather than the productivity during the design activity itself.

In evaluating design effectiveness, it must be recognized that the owner may have as great an influence on the final result as the design engineer. Among the impacts of the owners are proprietary technology inputs, design requirements or constraints, schedule or cost constraints, and changes imposed while the project is in progress. Furthermore, design effectiveness must be measured with recognition of the objectives of the owner, rather than against absolute or independent criteria.

The Construction Industry Institute has developed a method for systematically evaluating the effectiveness of a design project [CII 86]. This method has the flexibility to be useful in a wide variety of circumstances, including:

1. Widely differing types of projects
2. Mixtures of objective and subjective measurements of design effectiveness
3. Differing objectives and criteria for design effectiveness
4. Measurement of overall design effectiveness regardless of influence source, or measurement of designer performance when influences of designer and owner can be separately identified

9.4.2 Design evaluation

Design, including the activities associated with creative thinking, sophisticated engineering calculations, and translation of ideas into drawings, specifications, and procurement of major items for construction, occurs during the early stages of a project. Many contributors are involved during the design phase, including the owner, the various designer groups, vendors, and perhaps construction representatives. It is difficult then to evaluate the performance of a specific party, but much easier to evaluate the overall results of all parties.

Any evaluation method evaluates the total design effort, rather than the designer alone. Nonetheless, it can be used to establish goals and benchmarks for tracking selected performance during the design phase of a project and for evaluating performance against prescribed benchmarks.

Most owners, constructors, and designers agree that "design" encompasses the owner and contractor home-office functions of: (1) preliminary design; (2) project management (including cost estimating and control); (3) procurement; and (4) detailed design. None of the existing methods of design evaluation adequately determines the overall effectiveness of these diverse functions.

According to CII [CII 86], many factors are involved in evaluating a design. However, these can be grouped into main components of criteria for evaluation, weights for the various criteria, and a method for combining the various criteria and their weights into a single quantitative number for evaluation. The criteria and their weights are discussed in the following section. Methods for combining the criteria and weights into a single quantitative number are described using a technique called an *objectives matrix* [CII 86].

9.4.3 Design evaluation criteria

Evaluation criteria should be established at the beginning of the design process, understood and accepted by all concerned, and subsequently used to evaluate design effectiveness. The criteria should include only those elements that directly impact attainment of project objectives. Design performance goals should be reasonable and attainable, and achievements should be dependent primarily on the actions of those involved in the design process. Finally, the criteria should cover all major aspects of design responsibility.

The design-evaluation criteria depend on many variables. The type of project will affect the choice of the design outputs that are selected for measurement. The project schedule also can have an impact on the selection of criteria. Related to the schedule, the type of construction or design contract has a bearing on criteria selection. The design user must be taken into consideration when selecting criteria for evaluation of design. The owner, designer, and constructor have different interests in, and uses for, cost-effective design. It is important to keep in mind these interests when selecting criteria to evaluate design effectiveness. All design users, of course, are interested in completing the project on schedule, within budget, and with high quality.

The seven criteria most suited for an initial evaluation of design effectiveness are summarized in Table 9.1. Each criterion is essential to each design user, and can be evaluated during or immediately after completion of project construction. Not all of these criteria are easily

TABLE 9.1   Initial Design Evaluation Criteria

Criteria	Quantitative	Subjective
Accuracy of design documents	X	
Usability of design documents		X
Cost of design effort	X	
Constructability of design		X
Economy of design		X
Performance against schedule	X	
Ease of start-up	X	

quantified by measurable factors. Table 9.1 also shows which of these seven criteria are quantifiable or must be subjectively rated by personal judgment. Although subjective ratings are valuable, methods should be pursued to develop quantitative evaluations for all criteria.

These seven criteria are not all-inclusive for measuring design effectiveness. Other criteria relating to operability, maintainability, safety, plant operating efficiency, and plant performance are at least of equal importance. These latter criteria, however, require evaluations after a period of plant operation, and by different personnel than those involved in the design construction and start-up of a project. The seven criteria should be evaluated before the project team disperses and important data become unavailable. Detailed discussions follow for each of the seven criteria.

9.4.3.1 Accuracy of design documents. Because specifications and drawings are the most readily identifiable products of a design, they are an important measure of design effectiveness. The accuracy of the design-documents criterion measures the frequency and impact of changes in the drawings and specifications. These changes are related to document revisions, and entail extra work in both the design and construction phases.

9.4.3.2 Usability of design documents. This criterion determines the ease of use of the design documents by construction forces, and relates to the completeness and clarity of the drawings and specifications.

9.4.3.3 Cost of design. The cost of the design effort is the only criterion that actually occurs totally in the design phase. This can be quantified by the cost-effectiveness of the design activities compared to original (plus approved changes) budgeted amounts and overall project costs.

9.4.3.4 Constructability. Constructability is the optimal use of construction knowledge and experience in planning, engineering, procurement, and field operations to achieve overall project objectives. Implementation of a planned constructability program helps to optimize project costs by successfully integrating construction knowledge into design engineering.

9.4.3.5 Economy of design. The economy of design criterion relates to overdesign or inefficient design. Overdesign can be reflected by the amount of oversized members and overspecified materials. A poor physical layout of the facility can be an indication of inefficient design.

9.4.3.6 Performance against schedule. All design users agree that the proper scheduling of design documents and designer specified/procured materials significantly affects a project. The performance-against-schedule criterion reflects the timeliness of design document and materials delivery.

9.4.3.7 Ease of start-up. The ease of start-up is a partial indication of the accuracy and efficiency of the design. A measure of the efficiency is obtained by comparing planned to actual start-up time. The number of operators and maintenance personnel required during start-up can also be an indicator of the ease of start-up.

9.4.4 Criteria summary

The seven criteria discussed in the previous section are important to all design users, and are relevant regardless of the type of project, construction project activity, or impact of any project variable. Each criterion should be used in any initial evaluation of design effectiveness, but this list should not be construed as absolute or all-encompassing. Because of industry, company, or project differences, some design users may identify additional important criteria.

Not all of these seven criteria influence or indicate design effectiveness to the same degree. Several of the criteria may be better indicators of an effective design, and deserve greater emphasis during an evaluation. Criteria weighting is necessary to properly recognize these variations, and evaluations can be combined using an "objectives matrix" technique discussed in the following section.

9.4.5 The objectives matrix

Riggs [Riggs 85] and others have suggested the use of a technique called an objectives matrix for productivity evaluation. The same concept can be used to develop an effectiveness measurement for design.

An objective matrix consists of four main components: criteria, weights, performance scale, and performance index. The criteria define what is to be measured. The weights express the importance of the relative criteria to each other, and to the overall objective of the measurement. The performance scale compares the measured value of a criterion to a standard or selected benchmark value. Using these three components, the fourth component, the performance index, is calculated, and the result is used to indicate and track performance [Stull 86].

9.4.6 Concluding remarks on design-effectiveness evaluation

A design-evaluation matrix can be used for any project type or area of a project. The criteria and weights used in the matrix can be modified to fit any project, and the evaluation can adapt to most needs. It can be as simple (subjective ratings) or as sophisticated as desired. Quantitative measures are obviously preferred to subjective ratings as a means of evaluation. Considerable data from many types of projects are needed to develop norms for values of subcriteria.

Given the many complexities and variables of the total design process, no measurement system can yield absolute quantitative results that are applicable without interpretation to all design situations and circumstances. However, the method for evaluating design effectiveness can be used to

1. Develop a common understanding among the owner, designer, and constructor concerning the criteria by which design effectiveness on a given project will be measured

2. Compare design effectiveness of similar projects in a systematic and reasonably quantitative manner, highlighting performance trends

3. Identify opportunities to improve the effectiveness of the entire design process and contributions to the ultimate result of all participants

9.5 Summary

Design is a process with a substantial degree of subjectivity, and is influenced by many factors. Nonetheless, it provides a project's central point of translation of ideas into specific instructions.

The method presented herein is for the evaluation of a design, not a designer. It is recognized that many parties and factors influence the final design product. The proposed method is based upon the following seven evaluation criteria:

- Accuracy of design documents

- Usability of design documents
- Cost of design
- Constructability
- Economy of design
- Performance against schedule
- Ease of start-up

Not all of the criteria influence a design to the same degree. A method is presented for combining weights and performance ratings of the seven criteria into a single performance index. The method utilizes an objectives matrix approach and provides the flexibility of using a variety of criteria, weights, and measuring systems to compute the performance index.

The ability to measure design effectiveness using the proposed method represents an important step in a broader effort to improve the total design process. Such an effort encompasses identification of the effect on design effectiveness of: (1) various inputs to the design process; and (2) the systems and techniques employed by the designer.

References

[AASHTO 85] *AASHTO Guide for the Design of Pavement Structures,* Vol. 2, Appendix EE, American Association of State Highways and Transportation Officials, Washington, D.C., 1985.

[CII 86] The University of Texas at Austin, "Evaluation of Design Effectiveness," *A Report to The Construction Industry Institute, Publication 8-1,* July 1986.

[Chesher 85] A. Chesher, and R. Harrison, *User Benefits from Highway Improvements, Evidence from Developing Countries,* World Bank Report, Johns Hopkins University Press, 1985.

[Darter 71] M. I. Darter, "Uncertainty Associated with Predicting 18-Kip Equipment Single Axles for Texas Pavement Design Purposes," Center for Highway Research, The University of Texas at Austin, Oct. 1971.

[Darter 72] I. I. Darter, B. F. McCullough, and J. L. Brown, "Reliability Concepts Applied to the Texas Flexible Pavement Design System," *Highway Research Record 406,* Highway Research Board, Washington, D.C., 1972.

[Darter 73a] M. I. Darter, W. R. Hudson, and J. L. Brown, "Statistical Variations of Flexible Pavement Properties and Their Consideration in Design," *Proceedings, Association of Asphalt Paving Technologists,* 1973.

[Darter 73b] M. I. Darter, and W. R. Hudson, "Probabilistic Design Concepts Applied to Flexible Pavements System Design," *Research Report 123-18,* Center for Transportation Research, The University of Texas at Austin, 1973.

[Riggs 85] J. L. Riggs, "What's The Score?" *The Military Engineer,* Vol. 77, No. 50, Sept.-Oct. 1985, pp. 496–499.

[Stull 86] J. O. Stull, and R. L. Tucker, "Objectives Matrix Values for Evaluation of Design Effectiveness," *A Report to The Construction Industry Institute, The University of Texas at Austin,* Aug. 1986.

[Zaniewski 82] J. P. Zaniewski, G. Elkins, B. C. Butler, M. Paggi, G. Cunningham, and R. Machemehl, *Vehicle Operating Costs, Fuel Consumption, and Pavement Type and Condition Factors,* Federal Highway Administration, 1982.

10

Construction

10.1 Introduction

To fulfill its purpose, an Infrastructure Management System (IMS) must follow through from the design phase to the implementation phases of construction, maintenance, and rehabilitation, plus data feedback. The transition from design to construction represents one of the most important organizational boundaries in the management system.

Construction converts a design recommendation into a physical reality. Successful construction meets the planning and design objectives within budget and time constraints. It may be sufficient if, after bidding, contract award, construction schedule, materials supply, and processing, actual construction and quality control are conducted in a routine manner. It is not sufficient, however, to perform the design function without thought to construction or maintenance.

If conditions in the field differ from the design inputs or assumptions, changes have to be made. Such changes, at least major ones, should not be made without consulting the design group. Proper management will help ensure that sufficient communication occurs, that new or innovative design or construction solutions are not inhibited, and that as-constructed documentation is complete and understandable. This latter point is most important. Because of the usual hurry and day-to-day problems of construction, it is easy to delay or avoid such documentation as actually built. There have been many experiences in which a search of the files after construction yields incomplete or erroneous information. Proper feedback and plans for rehabilitation depend on reliable and complete as-constructed documentation.

This chapter does not attempt to provide details on construction practices, construction control, or construction management per se.

These are comprehensive subjects that are treated in a variety of construction manuals and guides published by various agencies. Attention in this chapter is focused on: (1) interrelationships of construction with other phases of infrastructure management; (2) constructability of the project; (3) construction quality assurance; and (4) documentation or data that construction should produce for construction use and the use of others. If these functions are carried out systematically, then the normal and expected variations in construction methods, equipment, materials, and environment, can be taken into account properly.

10.1.1 Construction documents

To move the selected design to the construction phase, a set of documents defining the details of the selected design is needed. In some cases, alternative designs may be offered for contract bidding and then the details of each are needed. These documents not only convey details for construction, but also serve as legal documents in procuring the services of a contractor or construction agency.

The traditional documents of design and construction are "plans and specifications." These usually consist of the following:

- A set of drawings that give detailed dimensions and other design aspects

- A set of project specifications that describe in detail the materials to be provided, their arrangement, required characteristics, and so on

- A set of "standards and specifications" that have been previously approved by the agency and are in general use

The first purpose of these implementation documents is to describe the proposed project to the construction group. For example, because a road project is usually built in conjunction with bridges, drainage facilities, and other items, the interrelationships involved also are described. In some agencies, the construction group is equipped to construct the project as designed, and in such cases construction can begin immediately. Usually, however, the construction group has a supervisory or control function and the actual construction is done by an independent agency under a legal contract. In these cases, the documents become highly important because they serve a second purpose as the basis for bidding, pricing, and agreements about payment for the work.

10.1.2 Construction management

Construction management involves the use of physical, financial, and personnel resources to convert designs to physical reality. This general concept intersects infrastructure management in the actual building of the facility. The processes are successful when stated planning and design objectives are met and the facility is put into service.

The process of construction management is comprehensive and complex. It involves many considerations, such as estimation, designation, and scheduling of activities; organizational and personnel aspects; legal aspects; finances and cost control; and records. There are many published books and manuals available on the subject. Although it is beyond the scope of this book to treat this subject in depth, the levels of construction management as they apply to public agencies are subsequently discussed.

10.2 Construction as Related to Other Phases of Management

An effective IMS depends on communication and coordination among all phases. General information flows to construction from the other plans of the management system. Details of these other phases have been discussed in previous chapters. Construction receives vital information from all of the phases.

Construction also provides information to these other management phases. The general nature of this set of information flows is given in Figure 10.1. The diagram shows that information from construction is vital to efficient management in the other phases.

Five phases, or subsystems, are considered in Figure 10.1: planning or programming, design, construction, evaluation, and maintenance. On first thought it might seem that construction should interface only with design and maintenance, but this is an incomplete formulation. Construction is important to the entire management system, and these relationships can be examined in terms of the information flows illustrated in the figure. These information flows occur in both directions—to and from the construction phase.

10.2.1 Planning and construction

The planning, or programming, phase provides the "what, when, and where" type of information to construction and other phases. Information on the type of work required and the location of the work can provide direction for both short- and long-term construction planning for work force and other resources. Often, this preconstruction information, involving sufficient lead time, can provide better contract prices based on better availability of contractors and less crowding of the work. It is also possible to accomplish the preconstruction activities of surveys, property acquisition, utilities relocations, and so on in a much more timely, economical, and efficient manner. The overall system, or network economic planning and analysis, will provide

Provided by Construction

PRE-PROJECT ENGINEERING	CONSTRUCTION ACTIVITIES	PROJECT FINALIZATION	POST-PROJECT
			Safety provisions (signs, markings)
Utilities	Construction progress	Final qualities and	
Materials acquisition	Actual quantities	locations	
and evaluation	Quality Control	Final costs	
Contract preparation	Actual properties	Records of special	
and tenders	and geometries	problems	
Project organization	Records		
Specifications	Actual costs		

PLANNING USE	DESIGN USE	EVALUATION USE	MAINTENANCE USE
Program Schedule updates	Evaluation of design assumptions,	Selection of evaluation sections	Scheduling and allocation of
Updates of project cost estimating procedure	materials estimates, cost estimates, analysis	Scheduling of measurements	maintenance resources
Updates of progress estimating procedure	methods, materials properties used, etc.	Relating measured performance and behavior to materials, environment, construction variations, etc.	Development of efficient maintenance procedures

Figure 10.1 Information provided by construction for potential use by other management phases.

information for scheduling contract tenders, personnel, and resources and provide for consideration of time restrictions and related items.

In the reverse direction, construction relates to programming in many ways. For example, comparisons of final contract costs and materials quantities with the original programming estimates can help in updating for future programming. Monitoring of construction in progress can point out the need for schedule adjustments. Cost overruns are not always available, and early knowledge can make it possible to reprogram without penalty to other contracts. Finally, completed construction and project openings are directly important to the next round of programming.

10.2.2 Design and construction

The plans and specifications provide direct design input into construction. This is the most obvious and direct interaction between the two phases. In addition to contract documents and details for construction, however, proper interaction in the early stages can assist construction in the preliminary project field-engineering phase and in setting up preliminary methods for locating materials and establishing quality- and quantity-control techniques.

Feedback from construction to design is equally vital. A seemingly economical design is not effective if it results in unusual or unmanageable construction problems. For example, a material may appear to be economical based on previous data, but a particular job or location may result in short supply and greater expense than expected. On the other hand, new construction techniques may lead to design changes. For example, all major highway departments previously required thin lifts of asphalt concrete in construction to gain adequate compaction. However, with the advent of heavier and variable-tire-pressure pneumatic rollers, and some construction trials, it has been shown that thick lifts may actually give better compaction under many situations. This type of feedback from construction led many agencies to change their specifications. A good management system can help in keeping new ideas from being discarded because of old methods.

10.2.3 Evaluation and construction

In its broadest sense, evaluation provides input to the construction phase. Continuous monitoring and evaluations of existing facilities can show that certain construction methods or technology, although acceptable under existing specifications, can lead to premature deterioration or rapid rates of deterioration. For example, aggregate gradation is controlled on a before-compaction basis. Subsequent testing in later evaluations may show that construction has actually degraded the material excessively, which may lead to structural weakness or drainage problems. A continuous and complete data-feedback system obtained by continuous evaluation of a structure and its components can assist in solving many such potential construction problems.

The input from construction to evaluation is direct and important. The as-built construction records provide the initial or zero-age evaluation of the facility. Furthermore, if the construction data are good and effectively recorded, they will form the backbone of all future evaluation because the data will be more detailed than can ever be obtained again. In addition, these records can be used effectively to assist in selecting the initial location of sections to be evaluated periodically. This is important because most evaluation is a sampling process.

During periodic evaluations, unusual problems observed should be checked directly against construction records. Unfortunately, in the past construction records have often been inadequate to fulfill these needs. A properly applied management system will assist in solving this problem.

10.2.4   Maintenance, rehabilitation, and construction

The last phase considered in relation to construction is maintenance and rehabilitation. Both routine repairs and work intended to restore or improve a facility is included in the definition of maintenance and rehabilitation. Because this phase is a continuing process, careful consideration of the maintenance required and where it will be done may show some patterns that provide feedback to construction and show up weaknesses in the quality of specifications, construction methods, and/or materials and supervision methods. Because restoration or rehabilitation work is often done by contract, it is considered by many to be a type of construction. In any event, scheduling of the work involved will require consideration and supervision by the construction group. Thus, it requires close cooperation and coordination.

Regarding the information flows from construction to maintenance, effective maintenance techniques and procedures are dependent on the actual materials and construction methods used. They are also related to problems that were encountered during construction.

For example, poor concrete performance can often be shown to be the result of a period of unusually hot weather encountered during construction, perhaps interrupting placement or disturbing finishing operations. In such cases, the as-built construction records form the keystone of the maintenance database. The information on final acceptance, claims, and adjustments can serve as the basis for the preliminary scheduling of maintenance programs and resources. Likewise, knowledge of the actual quantities and qualities of the materials used is essential to future maintenance planning.

10.3   Constructability

Knowing how to build facilities is crucial to optimal planning, designing, and administration of any construction. Know-how is the result of education and experience, and, like any other knowledge, once learned it must be extended and reviewed.

Construction experience is a premeditated learning process whereby a participant in, or observer of, construction operations understands what is taking place and can further describe and communicate to others the events, their purpose, and outcomes.

10.3.1   What is constructability?

Constructability has been defined by the Construction Industry Institute (CII) [CTR/CII 87] as the expediency with which a facility can be constructed. Constructability is enhanced by the proper use of construction knowledge and experience in planning, design, procurement, and field operations to achieve overall project objectives [CII 86].

Because constructability enhancement (CE) is by nature multidisciplinary and multifaceted, it means different things to the various participants in a project. To the project owner, constructability affords the opportunity on construction projects to achieve greater efficiency, with resulting lower cost, shortened schedule, or improved quality. To the designer, it is an understanding of the methods and constraints of the actual construction required to execute the design being made. To the contractor, it is a combination of the effort required to implement the design efficiently and the opportunity to minimize his or her effort and resource expenditure.

The Construction Industry Institute has identified a number of constructability concepts applicable to the different phases of a project. Briefly, these concepts address project execution planning, conceptual project planning, specifications, contracting strategies, schedules, and construction methods, including those concerning pre-assembly, site layouts, design configurations, accessibility, and adverse weather [CII 86]. Constructability improvement has been studied and applied to many segments of the industrial construction industry, and to highway construction and public facilities as well [De Vos 89, Fisher 89, O'Connor 86].

Constructability is indeed practiced to some extent by planners and engineers of state Departments of Transportation (DOT), although it may not have been formally defined and thought of as a primary factor in highway design and construction.

10.3.2 Relationship between constructability, value engineering, and productivity

Value engineering (VE), has similarities to constructability, and the differences may not be apparent at first. Innovative construction practices, leading to cost reductions, can be attributed to both constructability and VE, from which it might be concluded that the two are synonymous. There are similarities in objectives and result, but the scope and reach of constructability and VE are different, as shown in Figure 10.2.

Value engineering is defined as a disciplined procedure for analyzing the functional requirements of a product or service for the purpose of achieving the essential functions of the product or service at the lowest total cost. Total cost, in this case, takes into account the owner's cost of planning, design, procurement and contracting, construction, and maintenance over the life cycle of the product or service and may also consider user costs. In the case of a governmental

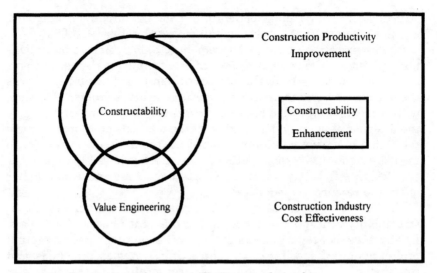

Figure 10.2   Construction industry cost-effectiveness relationships.

agency responsible for delivering a service, user costs should be considered in addition to production costs when the cost-effectiveness of a service is estimated.

Highway constructability, for example, is primarily concerned with optimal construction costs consistent with the function and quality requirements and boundaries set out by the standards or policies of the DOT. Thus, constructability enhancement (CE) should be considered, as one of several tools of VE. The effectiveness of both VE programs and CE programs is dependent upon participants who can contribute and work as members of a team.

From a practical viewpoint, contracted construction costs are by far the largest item in an agency's budget, and it is likely that the most significant improvements in productivity will come about through constructability enhancement. In the case of an agency which operates more or less on a fixed annual budget, productivity is equated with lower construction costs and/or improved quality, which translates into increased durability and improved operations. One of the functions of CE, therefore, is to provide feedback of construction experience to programming, planning, design, and construction.

The application of VE is similar in many respects to CE in that both concepts are applicable throughout the planning, design, and construction phases, and each is likely to have the most impact during the early stages of project development.

The foregoing discussion underscores the significance of early and effective consideration of constructability as part of an IMS. This is particularly true during the project planning and design phases, where the value of construction knowledge and experience, applied at the right time, can render the highest dividends. It is clear that once the project advances beyond these phases, investment and other commitments generally accumulate at rates depicted by the well-known S-curve. The ability to make changes to a project relates strongly to the S-curve. In the same vein, the ability to influence and control costs reduces rapidly, since it is inversely proportional to the same curve. These trends are shown in Figure 10.3.

Thus, in summary, the relationships among productivity, value engineering, and constructability, all of which are important to effective infrastructure management, are as follows:

- Productivity is a measure of the output/input ratio in constructing a facility.

- Value engineering is concerned with providing the required functions of the facility at the least cost.

- Constructability is a measure of ease or expediency of construction.

Figure 10.3 Significance of early decisions. (Adapted from [Azud 69, CTR/CII 89]).

10.3.3   Factors affecting constructability

During the development of the various elements of a project, constructability is influenced by numerous and diverse factors. A listing of these factors is given in Table 10.1, in which the factors have been grouped into seven categories for subsequent incorporation into a knowledge base.

With the assistance of an agency steering committee, a listing of constructability issues or concerns can be identified and prioritized for research and development, as outlined in Table 10.2. Understandably, the list may be long and the topics rather broad in scope.

TABLE 10.1   Factors Affecting Constructability (with Highway as an Example) [CTR/CII 89]

Environmental Systems	

■ Site
    ■ Topographical (including
      accessibility)
    ■ Geotechnical
    ■ Hydrological
	■ Infrastructural (including vehicular
	  traffic)
	■ Political/legal/regulatory
	■ Economical/sociological/financial
	■ Technological

Project scope	Information and communication

Operational requirements
Facility characteristics
    ■ Structural composition
        ■ Complexity
        ■ Scale
Financial and time constraints
    ■ Budget and schedule
	Documentation/transmission/interpretation
	    ■ Availability/source/accuracy
	    ■ Clarity and conciseness/completeness
	    ■ Consistency/compatibility/ambiguity
	    ■ Timeliness and frequency
	    ■ Relevancy

Resources	Processes/methods pertaining to

Material/manpower/plan and equipment
    ■ Availability
    ■ Variability/flexibility
    ■ Suitability
    ■ Intrinsic attributes
	Planning, design, specification, and
	  estimate
	    ■ Procurement/bidding
	    ■ Construction
	    ■ Maintenance

Controls	Innovation

Quality assurance/quality control;
testing and inspection
    ■ Cost and financial control
    ■ Schedule control
    ■ Productivity measurement
	Awareness of prompters; recognition of need;
	stimulation/encouragement
	    ■ Motivation and freedom to innovate
	    ■ Capability to innovate
	    ■ Resources and R&D
	    ■ Support of champion/innovative leaders

TABLE 10.2 Highway Constructability Issues and Concerns

- Planning and design guidelines for enhanced highway construction
- Specification improvements for enhanced highway constructability
- Selection, processing, and management of materials
- Constructability enhancement through innovation
- Facilitating construction under traffic
- Facilitating future expansion and upgrade
- Optimal utilization of plant and equipment
- Optimal risk/responsibility allocation
- Constructability program implementation

10.3.4 Constructability enhancement programs

To enhance constructability, good objectives must be established. Typically, they are there to:

- Increase productivity
- Reduce project costs
- Reduce project duration
- Reduce delays/meet schedules
- Eliminate unnecessary activity
- Reduce physical job stress
- Promote safety on construction sites
- Reduce conflict
- Increase quality

10.3.5 The highway constructability knowledge base

As defined by the Texas Department of Transportation (TxDOT) and the Center for Transportation Research [CTR/CII 89], a highway constructability knowledge base (KBS) is a collection of ideas obtained by personal interviews, expert sessions, and literature reviews. A computerized information-retrieval system for detailed treatment of constructability ideas is being developed [Redelinghuys 89]. This will relate the hierarchical diagrams to the various elements of highway projects to be constructed—for example, pavements. The hierarchical diagrams are also related to other aspects, such as the constructability factors in Table 10.1 and the applicable engineering phase. This approach also offers an efficient structure for continuing analyses and further research. Typical examples of applications or solutions to constructability constraints are given below:

- Acquire right-of-way (ROW) in a timely manner; resolve problematic ROW parcels early
- Minimize demands for ROW likely to be difficult to acquire; employ techniques for achieving steeper cuts/fills when necessary
- Reduce delays in utility adjustments; design to avoid utility adjustment by using utility bridges or tunnels
- Ease or secure the traffic/construction interface; use concrete safety barriers
- Speed up on-site bridge construction; optimize off-site prefabrication for bridge construction
- Provide space for contractor accessibility and staging; early consideration of additional ROW needs of constructor for storage and staging, access, and parking
- Employ tactics to reduce overall project duration; ensure efficient scheduling of construction activities; effective use of liquidated damages and incentives
- Optimize pavement unit productivity; minimize the number of pavement layers, particularly in intersection design
- Develop a pavement construction execution plan; develop a comprehensive inspection plan commensurate with the anticipated progress of the project
- Design earthworks to enhance constructability; allow innovative deep-lift compaction
- Facilitate future expansion and upgrade; locate utilities to minimize future adjustments; design to take into account future maintenance and expansion
- Minimize specification-related problems; change unrealistic tolerances; remove references to obsolete methods or materials; ensure consistent interpretation by removing ambiguity and educating and training users that the cost incurred in the design of the utility bridge to carry the weight of the signs and the additional wind load was negligible

10.3.6 Program implementation: barriers and recommendations

10.3.6.1 Barriers. Although the benefits of a Constructability Enhancement Program have been demonstrated, barriers to practicing good constructability are common. The Construction Industry Institute points out that managers should be aware of these barriers, listed in Table 10.3, and challenge them.

10.3.6.2 Recommendations. According to CII, the following recommendations should be considered in implementing a highway constructability program:

1. The commitment of senior management to constructability must be obtained.

2. A strong approach to project management with a single point of responsibility should be pursued.

3. Project-execution plans should be developed for large complex projects during a project-concept conference. Additional planning meetings and design reviews should involve greater participation of involved parties.

4. A proactive approach to constructability needs to be taken. Overreliance on late, reactive design reviews should be avoided.

TABLE 10.3 Barriers to Program Implementation

Barriers to communication and design-construct integration	Barriers to utilization of advanced construction technologies	Barriers to innovation
Contract time	Lack of awareness of technologies/inadequate communication	Discouragement of personal initiative, and perceived lack of freedom
Lack of time	Lack of necessary training	Failure to recognize opportunities
Lack of field feedback	Regulatory inhibitors	Lack of personal creative ability
Failure to document and communicate "lessons learned"	Institutional and individual resistance to change	Lack of tools
Lack of construction experience	Reluctance to deviate from current and proven standard operations	Lack of senior support of champions

5. Feedback from the field, if not forthcoming, should be solicited on a periodic basis, prior to, during, and after construction. This feedback should involve department personnel, contractors, and suppliers.

6. Post-mortems should be conducted upon completion of all projects. These should be attended by representatives from the owner and the contractor. Other interested parties should also be invited. These meetings should be used to report on "failures" as well as "successes." Increased opportunities for site visits should be made available.

7. Management training programs that promote communication and integration between design and construction should be conducted. Project "team building" should be initiated on a trial basis and should include exercises for developing team-leadership skills.

8. An accessible and current knowledge base of "lessons learned" should be maintained. Advanced, computerized systems are being developed for storing and retrieving the information.

Figure 10.4 illustrates the foregoing recommendations.

10.4   Construction Quality Control and Quality Assurance

Construction quality assurance is a complex and detailed process to guarantee that the finished facility is built to the standards desired by the owner (more precisely, as defined in the design documents). There are several aspects of quality assurance that are important,

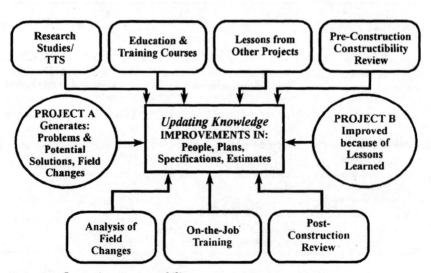

Figure 10.4   Improving constructability.

including quality control, quality assessment, and testing [Warden 63, Willenbrock 76]. Quality control, quality assurance, and total quality management are further discussed in Chapter 11. The discussion in this chapter is directed to construction.

No matter what type of quality assurance is undertaken, all of them start with specifications. There are several types of specifications used in the construction of infrastructure. These are variously termed as

1. Methods and materials specifications

2. Recipe specifications

3. End-result or end-product specifications

4. Performance-based specifications

5. Guarantee or warranty specifications for a specified time or utilization period

At the present date, methods and materials specifications are the most common. Although there is some difference in terminology, recipe specifications are generally the same as method and material specifications. That is, the recipe requires the inclusion of certain ingredients or materials and requires that they be mixed, manufactured, installed, or placed in a specified way.

10.4.1 Recipe specifications

The most typical specification in use today is the recipe specification. For Portland cement concrete, for example, the construction specification usually specifies aggregate type, and the gradation percent of aggregate, cement, and water that will be blended for the mix plus one or more additives to improve workability and increase set time. A mixing time may also be specified. Such specifications for concrete, however, also generally specify a minimum acceptable strength; for example, a breaking compressive strength of not less than 20,670 kPa (3,000 psi).

A look through any specification book for a city, county, or state public agency, as well as any major contract for building construction in the private sector, reveals that most of the specifications are of the recipe type. This represents a conflict in that the contractor is told everything to do in the process, and then what minimum strength must be obtained. On the other hand a liability exists if the recipe is faithfully followed, using the specified materials, but does not meet the strength requirement. In such a case, where does the fault lie?

This has been handled in the past by giving a sufficient factor of safety in the strength requirements, and in re-using good historical materials. Nevertheless, many legal claims must now be settled in court or by arbitration over recipe specifications with conflicting components. It is rather like telling one's partner exactly what to include in his or her cake, and then being angry because it does not suit your taste.

10.4.2 End-result specifications

There is an increasing use of end-result or end-product specifications to overcome some of the problems with recipe specifications. Rather than specifying the materials and methods to be used, the end-result specification concentrates only on the final product. That is, in our metaphor, the taste of the cake rather than how it was made. End-result specifications are now generally accepted by both contractors (as producers) and public agencies (as purchasers) of infrastructure products [ASTM 78, AGCA 77, Gappinger 85]. There are several reasons for the increasing popularity of end-result or product specifications in contract work. First, end-result specifications help to determine the responsibility for product quality control and quality assurance. Secondly, they provide economic benefits to contractors who produce high-quality work and to the purchaser (public agency) in the form of improved products and reduced life-cycle cost. This occurs because end-product specifications allow the contractor or producer the flexibility to use innovation in producing the required end-results.

The use of end-result specifications is, however, a fairly recent phenomenon. In the last decade, for example, concern over ride quality on new pavement surfaces has prompted some state DOTs to develop end-result smoothness specifications [Harrison 89, Hudson 92, SDHPT 82, SDHPT 91].

10.4.3 End-result illustration

To apply an end-result specification, it is necessary to have a valid measuring tool that both the producer and owner can accept as understandable and valid and that can be checked and calibrated efficiently as needed.

In terms of pavement smoothness, a proposed specification has been developed by TxDOT [Hudson 92, TxDOT 91]. The main limitation to developing this specification was finding available roughness-measuring equipment that would provide a stable reference for ensuring contractor compliance. Several devices were examined, and a modification of the profilograph, developed in California [AASHTO 87], was selected because it best fulfilled the needs of end-result specifications, even though it was not the most widely used roughness-measuring device available. In this case, a device that produced a visual trace that could be seen by the producer and the buyer, as well as a stable reference

TABLE 10.4   Preliminary Smoothness Pay-adjustment
Schedule. [Hudson 92].

Profile Index (PI) inches/mile per each 0.1 mile section	Bonus or deduction percent of unit bid price
3.0 or less	+5
3.1 thru 4.0	+4
4.1 thru 5.0	+3
5.1 thru 6.0	+2
6.1 thru 7.0	+1
7.1 thru 10.0	+0
10.1 thru 11.0	−2
11.1 thru 12.0	−4
12.1 thru 13.0	−6
13.1 thru 14.0	−8
14.1 thru 15.0	−10
Over 15.0	Corrective work required

platform, were more important than speed of operation or ultimate
units used.

Any end-results specification must be applied realistically. That is,
the acceptable level must not be set so high that it is unattainable by
a good contractor. Conversely, it is undesirable to set the specification
acceptance criteria, or level, so low that any contractor can meet it. To
follow the principle of setting a realistic inducement, it is necessary to
obtain adequate field data on the general performance of the various
contractors carrying out the proposed procedures. Usually, the level
should be one that 10 to 20 percent of the contractors can typically
meet, and 80 to 90 percent will fail to meet. This means that the level
is obtainable, but a large number of existing contractors will need to
improve their procedures to meet it. Table 10.4 illustrates the specification produced for TxDOT.

The specification level also indicates another major principle of
good end-result specifications—bonus provisions. Experience shows
that contractors subject to being penalized feel much better if they
also have the possibility of being rewarded. When such a bonus, or
reward, system is included, end-results specifications are much more
likely to be successful.

10.4.4   Factors influencing evaluation and enforcement of a specification

In evaluating pavement smoothness, as with any other end-result
specification, it is desirable to have a standard and accurate instrument against which all other instruments can be compared.
Unfortunately, many times no standard instrument exists. Lacking
an acceptable standard, the next-best thing is to select a particular

instrument type and accept it as the standard output device for use in enforcing and developing the specification.

Repeatability of the measurements of compliance must be carefully evaluated before equipment and specification levels are set. The variability introduced by the equipment operator must also be evaluated. It should be remembered that these comparisons must be made under field conditions, that is, the conditions that are expected for acceptance and routine use in the field.

It is also important to know how variability could affect management interpretation by the same reader. That is, might a profilogram, for example, be interpreted differently in the morning (when the reader is alert), than it would be later in the day (after the reader has begun to experience fatigue)? In attempting to answer this important question, the study team in the TxDOT study collected information on variability considered to be realistic to determine what kind of variation can be expected in real-life data collection. This avoids the problem of selecting a band so narrow in the specifications that it is unreasonable to expect the contractor to meet it.

The data for this analysis were obtained by having the same interpreter read several profilograms several times for both high- and low-profile level pavements.

10.4.5 Summary of example

A new smoothness specification for flexible pavements, based on end-result criteria, was developed and implemented successfully in Texas (see Table 10.5). The development of this specification has led to the following recommendations.

Although most asphalt concrete pavements are relatively smooth, an improved specification shows that they can be made even smoother by paying a bonus to the contractor for high-quality work. End-result smoothness specifications are successful in reducing the small amount of roughness present in new asphalt-concrete pavement contracts. The bonus needs to be sufficiently attractive to induce consistent, high-quality work throughout the entire contract; however, because roughness in new contracts is relatively localized, the total bonus as a percentage of the bid price can be quite small. A highway department,

TABLE 10.5 Smoothness Specification for Flexible Pavements

PI (in/mi)	Percent of unit contract price
PI ≤ 9.0	100
9.0 < PI ≤ 10.9	99
10.9 < PI	Corrective work required

therefore, risks only a small increase in job price for the considerable benefits that accompany a rise in ride quality and serviceability indices. The bonus incentive has an additional feature related to contractor bidding: end-result smoothness specifications make it more difficult for inferior contractors to win bids. This is because experienced contractors familiar with smoothness specifications can discount their bonus into the bid price. In this way, the department does not pay a large total price for the contract. The bid plus bonus becomes very close to original bid-price levels. The Texas experience shows that if the contractor is inferior or inexperienced, problems with ride quality increase. End-result specifications, therefore, reduce the probability of poorer contractors being awarded pavement work.

In summary, final acceptance smoothness specifications for newly constructed flexible pavements and newly overlaid flexible pavements are excellent example applications of end-result criteria in highway construction. An end-result smoothness specification improves overall pavement ride quality, gives a better service to the road user, and provides the highway agency with an excellent means of specifying and assuring adequate quality of construction. Total bid plus bonus price for pavement work need not be much higher than traditional bids, and inferior contractors have a lower probability of winning bids. The Texas study suggests that end-result specifications are a highly desirable modification to the traditional smoothness specification [Hudson 92].

10.4.6 Performance-based specification

Performance-based specifications are substantially more complicated to develop and use than end-result specifications. The concept is to write a specification for the contractor or the facility provider that will require the entity constructing the facility to provide a given level of performance.

Performance is a time-history function. For example "a pavement shall retain a serviceability index of greater than 3.0 (on a scale of 5 to 0) for five years or more." Such a specification requires a history basis for defining the performance, and a measure of performance to be used for control or compliance. In some cases, performance is defined in the short term. One example is a high-performance car or boat that is "capable of reaching 96 km per hour (60 miles per hour) from a standing start in 6.0 seconds." Another example is a high-altitude airplane that is "capable of performing flight at an altitude of 12,000 m (40,000 feet)." These definitions of performance are not generally applicable to infrastructure management. In this case, performance is time-related, as outlined for the pavement, such that a building, bridge, water line, or sewer "shall serve at an acceptable level of service for a known period of time."

A good illustration of the problem associated with such specifications is the Strategic Highway Research Program (SHRP). In 1987, the SHRP was set up to develop performance-based asphalt pavement specifications. The research program set out to undertake all sorts of detailed studies of asphalt aggregates and mixes over a five-year program. It was clear, however, from the start that there was not time within the program to develop new materials, build pavements out of them, and truly observe the performance. Therefore, it was patently impossible to fulfill the stated objectives. The program management suddenly changed the title from performance-*based* specifications to performance-*related* specifications to indicate that the specifications were to include factors that were, in general, at least known to impact the long-term performance of the material. The fact that this was already a part of most existing specifications was perhaps not recognized by the SHRP administrative staff. It should be remembered that performance-based specifications cannot be validated until actual performance has been observed in the field and compared to the performance obtained with different qualities or types of materials.

Infrastructure management systems provide the ideal framework for developing performance-based specifications; however, it will take a number of years before such specifications can be developed based upon analysis of feedback data from an IMS database. Users of IMS are urged to close the loop between observed performance and the quality and reliability of construction to form a basis of future development of true performance-based specifications.

10.4.7 Guarantee or warranty specifications for specified period or utilization

Warranty specifications have not been widely used in infrastructure. They are, however, broadly used in areas such as the automotive industry. Improvements in the quality of automobiles have allowed manufacturers to increase from a one-year or 16,000 km (10,000 mi) warranty to a five-year 80,000 km (50,000 mi) warranty. Similar mechanical warranties are provided for appliances and other consumer machines, even airplanes and ships. Such specifications and warranties, while slow in coming even in the automobile industry, where millions of vehicles are manufactured and sold annually, are even harder to develop in civil infrastructure. One reason is that the automobile industry's very large size makes it relatively easy to obtain good experience and information on reliability, not only of the automobiles but of the various component subsystems, such as engines, transmissions, and brakes.

These warranties do not come free, and they clearly involve an increase in vehicle price. It is possible to estimate the relative cost of such warranties for automobiles by pricing options made available by the manufacturer for so-called extended warranties. A new automobile purchased by one of the authors came with the option of extending the warranty period for an additional 48,000 km (30,000 mi) by paying $750. The price for extending the warranty to 160,000 km (100,000 mi) would have been $1,500, bumper to bumper, all physical items covered. The relative higher unit cost for the longer extension (which amounted to about 3 to 5 percent of the cost of the automobile) accounts for the added wear and the additional chance of failure over the longer period.

The cost of such warranty specifications for highways, bridges, and buildings would likely be considerably higher than 3 to 5 percent because of the lack of broad-gauged reliability experience and because each construction facility varies from every other facility much more than individual cars vary in the automobile production line.

10.4.7.1 Applications of warranties to infrastructure. Since each infrastructure unit, such as a pavement, bridge, building, water line, or sewer line is unique, the guarantee approach is perhaps a better contracting and pricing mechanism than a direct warranty. This approach is often referred to as "privatization of infrastructure," and it has become somewhat more popular in the last few years. It provides the ability to fix the price of the use of the facility for the owner and/or user, but it does not serve in exactly the same role as a recipe or end-product specification. Nevertheless, it fits into construction quality assurance in the sense that if warranty specifications are used it is no longer up to the owner to inspect, test, and check the construction process on a day-to-day basis.

With warranty specifications, the type of testing for recipe or performance-based specifications is not necessary. However, some way of evaluating the satisfaction of the owner or the facility user over time is needed. In the case of the automobile this is fairly direct, but in the case of a highway or bridge it is more difficult. It is hard to say if the quality of the road serving the user has declined below an acceptable level unless some testing is done during the performance period. One measure might be the number of complaints from users that the ride on the road is rough, that the water pressure is low, and so forth.

10.4.7.2 Privatization of maintenance. One major approach to warranty specifications for existing infrastructure is the so-called "privatization of roadway maintenance," as further discussed in Chapters 11 and 13. This approach has been used in a number of countries, such as Argentina and Australia, for roads. It has involved designation of an entire area or network of roads to be bid on by a consortium of financiers, engineers, and contractors. In the State of New South Wales in Australia, for example, each consortium submits a bid in which they agree to provide necessary maintenance, reconstruction, and rehabilitation for a fixed period (e.g., five or ten years) and to do everything necessary to provide the facility at an acceptable service level for the contract price. All of the bids or tenders are then evaluated and a winning group is selected. The group may not necessarily be the lowest price tender, since there may be differences in the level of service and quality provided within the activity of different groups. These approaches have not been in place long enough to fully understand how well they are functioning. The concept is a good one, allowing for economies of scale, good private management, savings in social cost for government workers, and the benefit from the widely accepted concept of higher productivity obtained from private industry. This is the conventional wisdom, which is not necessarily the opinion of the authors and certainly open to debate.

10.4.7.3 Privatization of new facilities.   New facilities are easier to privatize. In the case of roads, this is being done for toll roads in Mexico, Spain, Canada, and other countries. It involves solicitation of offers saying that a roadway system is needed from point A to point B (usually from one major city to another), and it shall have certain connections, certain speeds, certain capacity, and so forth. Proposals are then received from various groups interested in providing a turnkey facility for a monthly or annual cost basis. In this case, the group also includes a bank or financial institution that arranges the necessary financing. Proposals are evaluated and one is selected—not always on the basis of minimum cost—on an overall effectiveness evaluation that is intended to provide the greatest benefit-cost ratio. In some cases, the costs are set as a specific toll for each vehicle using the facility, and the successful bidders are not allowed to charge more than the bid tolls. In other cases the annual cost is set, and the difference between collected tolls and the cost basis are borne by the owner or tolls are raised to cover the difference. In this case all costs, including interest, administration, maintenance, rehabilitation, and operation, are covered in the pricing.

The intention of these initiatives is to take full advantage of the efficiencies attributed to private-sector economics and entrepreneurship. There is inadequate history, however, to determine whether or not they are an effective mechanism for warranty specifications. In at least one reported case in Spain, the group winning the bid went bankrupt within three years. It is understood that the additional cost fell back on the government, even though there were warranty bonds and other mechanisms intended to cover this contingency.

10.4.7.4 Warranty specifications in shopping centers. The private sector offers a better model of warranty specifications, and one that has proven to be successful. It is the standard of the industry, and it is being widely used by major corporations that are capable of providing their own infrastructure but choose to obtain their infrastructure from construction and management specialists at a fixed bid price.

It is the current practice, for example, of many large retail agencies, such as Sears and Albertson's grocery stores, not to construct their own stores, but to become the so-called *anchor tenant* in a major shopping center. In such cases, the developer agrees to provide retail space, parking, utilities, and maintenance, required by the major tenant at a fixed or agreed-upon price, usually with inflation clauses and scales to cover unforeseen increases in utilities. Such contracts specify the quality and type of parking, air conditioning, floor coverings, lighting, and so forth, that will be provided for the contract price. In return, the tenant obtains needed infrastructure at a known price without becoming an infrastructure provider. As a result the tenants can concentrate on their primary business of retailing, without the need to extend themselves into unknown areas of construction, maintenance, and repair.

This approach is not a panacea that solves all problems, and there can still be disagreements as to whether or not the service contracted for is being provided adequately. For example, the authors were involved as consultants in a major lawsuit at a large shopping center in San Antonio, Texas. In this case, key anchor tenants had a strong disagreement with the shopping center developer about the quality of parking provided and about drainage problems that caused water to pond in front of their stores, making it inconvenient for the customers, who complained avidly. The approach, however, is a valid one, and infrastructure management systems can assist in ensuring that the process works effectively. Infrastructure management for users and providers should consider this approach as an option.

10.5 Summary

Construction converts a design into a physical reality. To do this successfully, the design must be well documented and "constructable." The actual construction must be well managed, and the work itself must undergo rigorous quality control and quality assurance. The latter can involve a variety of specification approaches, including methods, recipe, end-result, performance-based, and long-term warranties. Each has advantages and disadvantages, with the recipe approach relying on an extensive base of experience. End-result and end-product specifications are, however, finding increasing use in infrastructure management, and warranty-type specifications also have been included in a number of countries.

References

[AASHTO 87]   *Survey of Results of the 1984 AASHTO Rideability Survey,* AASHTO, Washington, D.C., 1987.

[AGCA 77]   *Statistically Oriented End-Result Specifications,* Associated General Contractors of America, Aug. 1977.

[ASTM 78]   "Quality Assurance in Pavement Construction," *Special Technical Publication 709,* ASTM, Philadelphia, Pa., 1978.

[Azud 69]   G. Azud, "Owner Can Control Cost," *Transactions of the American Association of Cost Engineers,* 1969.

[CII 86]   *Constructability: A Primer,* Construction Industry Institute, Austin, Tex., 1986, pp. 2–5.

[CTR/CII 87]   *Guidelines for Implementing a Constructability Program,* Center for Transportation Reserach, Construction Industry Institute, Austin, Tex., 1987, p. 10.

[CTR/CII 89]   F. Hugo, J. T. O'Connor and W. Ward, *Highway Constructability Guide,* Texas State Department of Highways and Public Transportation, The Center for Transportation Research, Austin, Tex., July, 1990.

[De Vos 89]   J. De Vos, "A Strategy for the Implementation of a Constructability Improvement Program in SASTECH," Master of Engineering thesis, University of Stellenbosch, Republic of South Africa, 1989, p. 5.

[Fisher 89]   D. J. Fisher, "Piping Erection Constructability Issues in a Semi-Automated Environment," thesis, The University of Texas at Austin, 1989, p. 28.

[Gappinger 85]   E. P. Gappinger, "End-Product Specifications: State of the Art," *Proceedings, 34th Annual Arizona Conference,* University of Arizona, 1985.

[Harrison 89]   R. Harrison, C. Bertrand, and W. R. Hudson, "Measuring the Smoothness of Newly Constructed Concrete Pavement for Acceptance Specifications," *Proceedings of the 4th International Conference on Concrete Pavement Design and Rehabilitation,* Purdue University, Ind., 1989.

[Hudson 92]   W. R. Hudson, T. Dossey, R. Harrison, and G. D. Goulias, *End-Result Smoothness Specifications for Acceptance of Asphalt Concrete Pavements,* Center for Transportation Research, The University of Texas at Austin, Jan. 1992.

[O'Connor 86]   J. T. O'Connor and R. L. Tucker, "Industrial Constructability Improvement," *Journal of Construction Engineering and Management,* Vol. 112, No. 1, 1986, p. 69.

[Redelinghuys 89]   J. Redelinghuys, "A Knowledge Base System to Enhance Highway Constructability," Master of Engineering thesis, University of Stellenbosch, Republic of South Africa, 1989.

[SDHPT 82]   *Statistical Specifications for Construction of Highways, Streets, and Bridges,* Texas State Department of Highways and Public Transportation (SDHPT), Austin, Tex., 1982.

[SDHPT 91]   *Ride Quality for Pavement Surfaces,* Revised 2-26-91, Texas State Department of Highways and Public Transportation, Austin, Tex., 1991.

[TxDOT 91]   *Operation of the California Profilograph and Evaluation of Profiles,* Texas State Department of Highways and Public Transportation, Materials and Tests Division, Test Method TEX1000S, Draft, 1991.

[Warden 63]   W. B. Warden and L. D. Sandvig, "Tolerance and Variations of Highway Materials from Specification Limits," *Special Technical Publication 362,* American Society of Testing and Materials, Philadelphia, Pa., 1963.

[Willenbrock 76]   J. H. Willenbrock, *A Manual for Statistical Quality Control of Highway Construction,* Federal Highway Administration, Washington, D.C., 1976.

11

Maintenance, Rehabilitation and Reconstruction (M,R&R) Strategies, Including Operations

11.1 Introduction

After construction, the next steps in the management of any type of infrastructure are maintenance and operations. In practice, there is a wide divergence in the definition and understanding of maintenance, particularly in the dividing line between maintenance and rehabilitation. In building infrastructure, for example, terms like *remodeling, restoration,* and *renovation* are commonly used rather than *rehabilitation.*

In highways and bridges, the dividing line between maintenance and rehabilitation is blurred by practice. Often it is divided along organizational lines involving a design division, a maintenance division, and a construction division. Sometimes, the definition depends upon funding sources. Historically, there was no federal funding in the U.S. for pavement and bridge maintenance. Maintenance was therefore repair work, done with local funds, while rehabilitation was work done with federal funds.

Some organizations originally defined maintenance and rehabilitation in terms of how the work was accomplished. If it was done by in-house forces, it was called "maintenance." If it was done by contract, it was termed "rehabilitation." These issues have changed, however, over the years. For example, federal funding in the U.S. is now available for highway and bridge maintenance as well as rehabilitation.

Additionally, a growing trend to privatization is blurring the dividing line between maintenance and rehabilitation.

Since there are so many types of infrastructure, this book deals with functional, or global, issues associated with maintenance and rehabilitation definitions. However, concise and relatively precise definitions for maintenance and rehabilitation will be provided.

In one sense, maintenance is perceived to involve routine items like changing the oil, replacing lightbulbs, filling potholes, repairing cracks, and associated activities, depending on the type of facility involved.

In another sense, however, maintenance is poorly understood. Remembering the systems concepts and methodologies discussed earlier, two important steps in the process are: (1) problem recognition; and (2) problem definition. To truly understand maintenance and be able to apply the understanding to improved infrastructure management, clear recognition and clear definition of maintenance in a functional sense is essential.

11.2   Definitions

To better define maintenance for infrastructure management, it is useful to first look at other industries and at various literature related to civil-engineering infrastructure. In fact a lot of work has been done, particularly in the space and aircraft industries. Areas of civil-engineering infrastructure, such as water and waste-water facilities, that involve both operations and maintenance also have applicable experience.

Examination of the literature from a range of sources shows a well-developed concept in reliability-centered maintenance (RCM). This RCM concept is used in military and civilian aircraft and space activities, where it was originally developed [King 86]. It plays a vital role in the safety of flying, where failure often means death. Later in this chapter, we will discuss reliability-centered maintenance in more detail, but the definitions presented here take into consideration that concept.

In the RCM concept, rehabilitation is simply not mentioned. The entire range of maintenance/rehabilitation activities, and even reconstruction and replacement, all fall under a maintenance definition. This is not satisfactory for all infrastructure, but the portion of the concept and the definitions that assist in our objectives can be used.

11.2.1   Terms for defining maintenance

Terms such as *routine maintenance, corrective maintenance, preventive maintenance, proactive maintenance,* and *reactive maintenance* are commonly used in practice. Other terminology is also used, such as *hard-time replacement, on-condition maintenance, condition monitoring, servicing task, rework task* (repair, overhaul, rebuild), *replacement task,* and *time-directed* (versus condition-directed) *activities.*

Nearly all sources recognize the need for some regular maintenance activities. An example is changing the oil in your car on a regular basis. This is both preventive maintenance and, if done regularly, routine maintenance. The term routine maintenance, however, also applies to time-based maintenance; for example, a field crew travels over every bridge within its service area to fill potholes, clean scuppers, and patch approach slabs. Preventive maintenance is done in an effort to forestall trouble and is often done routinely. However, not all "routine maintenance" is preventive in nature. Corrective maintenance is done to remedy an obvious damage or flaw, and is an after-the-fact activity. Thus, it may also be called *reactive* maintenance.

The term proactive maintenance is often applied to work carried out at the discretion of the maintenance crew, or as planned in an IMS, to prevent impending deterioration or failure. Preventive maintenance and proactive maintenance are similar.

Condition monitoring leads to on-condition maintenance (OCM). In this activity, regular inspections are made to monitor the condition of the infrastructure, and the maintenance is scheduled as a function of that condition. In the aerospace industry, oil changes, tire rotations, and so forth, are done as a servicing task on a hard-time replacement (HTR) basis. Rework tasks, such as engine replacement, may also be done as HTR. After several cycles of rework, old components are totally replaced with new ones. These concepts do not generally apply to pavements or bridges, and apply only partially to water works, sewage works, and other types of public infrastructure.

We will then limit our definitions to two basic divisions: proactive or preventive maintenance, and reactive or corrective maintenance. Rehabilitation will be defined later in this chapter.

According to the dictionary, maintenance is defined as "the action of keeping in effective condition, in working order, in repair." A more detailed functional definition suggests that maintenance deals with the specific procedures, tasks, instructions, personnel, qualifications, equipment, and resources needed to satisfy the maintainability requirement within a specific use environment. For our purposes a good, concise definition is as follows: "Maintenance is that set of activities required to keep a component, system, infrastructure asset, or facility functioning as it was originally designed and constructed to function." Within this definition of maintenance, a number of subsets can be defined.

Preventive Maintenance (proactive maintenance) is performed to retard or prevent deterioration or failure of a component or system.

Corrective Maintenance (reactive maintenance) is performed to repair damage and/or to restore infrastructure facilities to satisfactory operation, or function, after failure.

Routine Maintenance is any maintenance done on a regular basis or schedule. It is generally preventative in nature, but may be corrective.

Hard-Time Replacement (HTR) is replacement after a certain length of time, regardless of whether the component has failed or not. It is therefore a type of routine maintenance, but may also be corrective or preventive.

On-Condition Maintenance (OCM) is maintenance done in response to condition-monitoring actions indicating impending deterioration or failure. By definition it is a type of preventive maintenance.

Critical Maintenance is defined by some as the maintenance that must be done immediately to prevent imminent collapse or functional failure. An example might be a bridge with a fractured steel girder, or an abutment that has been hit and fractured by a passing barge or ship.

The relationships of the above subsets are shown in Figure 11.1. A major distinction between corrective and preventive maintenance is whether or not failure or damage has already occurred. In many cases, no distinct failure point can be identified; thus, the dividing line between the two is shown as a gray area.

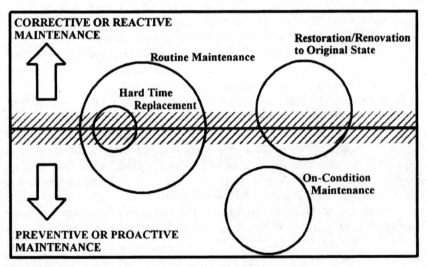

Figure 11.1 The relationships between subsets of maintenance.

11.2.2 Maintenance standards for historic properties

The U.S. Department of Interior has established standards for treating and preserving historic properties that have been used by the National Park Service since the mid-1970s. The four treatment approaches based on these standards emphasize the retention and repair of all historic fabric, as briefly described in the following sections [Weeks 96].

11.2.2.1 Preservation.

Preservation is defined as "the act or process of applying measures necessary to sustain the existing form, integrity, and materials of a historic property." Key ideas relating to preservation standards include the following:

- Use the property as it was used historically, or find a new use that maximizes retention of distinctive features.
- Preserve the historic character (continuum of property's history).
- Stabilize, consolidate, and conserve existing historic materials.
- Replace minimum amount of fabric necessary, and in kind (match materials).

11.2.2.2 Rehabilitation.

Rehabilitation is defined as "the act or process of making possible a compatible use for a property through repair, alterations, and additions, while preserving those portions or features that convey its historical, cultural, or architectural values." Key ideas relating to rehabilitation standards include the following:

- Use the property as it was used historically, or find a new use that minimizes retention of distinctive features.
- Preserve the historic character (continuum of property's history). Do not make changes that falsify the historical development.
- Repair deteriorated features. Replace a severely deteriorated feature with a matching feature (substitute materials may be used).
- New additions and alterations should not destroy historic materials or character. New work should be differentiated from the old, yet remain compatible with it.

11.2.2.3 Restoration.

Restoration is defined as "the act or process of accurately depicting the form, features, and character of a property as it appeared at a particular period of time." This is accomplished by removing features from other periods in its history and reconstructing missing features from the restoration period. Key ideas relating to restoration standards include the following:

- Use the property as it was historically, or find a new use that reflects the property's restoration period.

- Stabilize, consolidate, and conserve features from the restoration period.

- Replace a severely deteriorated feature with a matching feature (substitute materials may be used).

- Replace missing features from the restoration period based on documentation and physical evidence. Do not make changes that mix periods and falsify history.

- Do not execute a design that was never built.

11.2.2.4 Reconstruction. Reconstruction is defined as "the act or process of depicting, by means of new construction, the form, features, and detailing of a nonsurviving site, landscape, building, structure, or object, for the purpose of replicating its appearance at a specific period of time and in its historic location." Key ideas relating to reconstruction standards include the following:

- Do not reconstruct vanished portions of a property unless the reconstruction is essential to public understanding.

- Reconstruct based on documentary and physical evidence.

- Precede reconstruction with a thorough archaeological investigation.

- Preserve any remaining historic features.

- Recreate the appearance of the property (substitute materials may be used).

- Identify the reconstructed property as a contemporary recreation.

- Do not execute a design that was never built.

11.3 Maintainability

Maintainability has become an important concept in infrastructure management. For example, urban freeways and bridges often carry more than 200,000 vehicles per day; as a result it is very difficult to close such a facility, even for a short time at night. Ideal infrastructure management would balance the high user costs during maintenance shutdowns against the additional costs required to keep such facilities operating without closure for maintenance. This analysis involves the costs and benefits of maintainability.

Maintainability is defined in different ways in the cited literature. The basic subjective definitions are simple statements like "the ease with which a facility or system can be maintained" or "the reciprocal of the mean net time required to perform maintenance." Another example is "the capacity to carry out maintenance with ease and minimum expenditure while maintaining the safety of the crew and a desired level of accuracy in the repair work."

Maintainability is also a design element that considers the ease of maintenance in the future. Explicit interaction between maintenance personnel and designers in the design stage can significantly improve the maintainability of the design in terms of minimum costs with minimum environmental impact and a minimum expenditure of resources.

Maintainability often relates to access or "access areas." This involves being able to get to the area for servicing or repair. It is an important consideration for facilities that have mechanical and electrical equipment, such as waste-water and water-treatment plants, and hydroelectric dams where spacing and location of equipment elements are important. It can also relate to bridges. The Connecticut bridge panel that fell out several years ago, for instance, resulted from the corrosion of a pin connection that was covered with a plate, which made it not easily accessible for inspection, maintenance, or replacement.

Maintainability is, therefore, an inherent characteristic of a design. Blanchard et al. [Blanchard 95] define maintainability in a more objective way as:

> A characteristic of design and installation which is expressed as the probability that an item will be retained in or restored to a specified condition within a given period of time, when maintenance is performed in accordance with prescribed procedures and resources.

11.3.1 Measures of maintainability and availability

The following points must be considered for maintainability:

- If maintainability is to be an inherent characteristic of design, it must involve the designer's vision of the end product. However, it is obvious that the as-built end product is not completely in the hands of the designer. Therefore, true maintainability characteristics may be different to those perceived in the design.

- A fundamental question is whether maintainability is related to the frequency of maintenance actions performed during a system's life cycle. In one sense we might say maintainability should be simply expressed in terms of the mean time to repair (MTTR) [Smith 93]. However, this does not capture the probability aspect over a given period of time as defined by Blanchard et al. [Blanchard 95].

The design features on which maintainability depends are so varied, and their interrelationships are so complex, that no single all-encompassing indicator can be determined as representing the desired maintainability characteristics for a given system. The following measures of maintainability are reported by the U.S. Department of Defense report [DOD 66].

11.3.1.1 Qualitative maintainability. As a requirement, this would be a general statement to be incorporated in a system-design process. Examples are minimizing complexity, accessibility of components and parts, and built-in self-test features. These qualitative statements become maintainability objectives for the design process.

11.3.1.2 Quantitative maintainability. A quantitative maintainability requirement is a definite statement of the allowable resources or time required to perform a given type of task. The measure of maintainability works in parallel with the availability of the system to perform specific functions, as discussed later in this chapter. The mean time to repair (MTTR) is often selected as a measure of maintainability, as follows. *MTTR* (mean time repair) is the mean of the times required to repair an item or system during a given period. Other important parameters used in maintainability and availability are *MDT* (mean down time), which is the total inactive time for the equipment during a given period, and *MTBF* (mean time between failures), which is the mean of the total active time for the component during a given period.

11.3.2 Maintainability analysis

Maintainability analysis precisely defines a system's maintainability requirements and provides the designers and the organizations responsible for the system's design and development with timely information as to how effectively the established repair-time goals are being met.

The first step in conducting a maintainability analysis is to establish maintainability controls in the form of qualitative and quantitative requirements. To determine the quantitative maintainability requirements for each subsystem, the repair time of the subsystems must be budgeted so that the mean is less than the allowable MTTR for the system.

Since MTTR for a system is its principal maintainability requirements, it is essential that means be developed early in a program for

allocating and controlling the MTTR of each subsystem. This allocation is managed by determining the contribution of active downtime
for each subsystem to the total system downtime, and evaluating
these contributions against the established MTTR for the system.

11.3.3 Availability concept

Availability is the probability of a system or component being in service when required. Inherent availability is given by the following
expression:

$$\text{Availability} = \frac{\text{MTBF}}{\text{MTBF} + \text{MTTR}} \qquad (11.1)$$

Provision of availability is a major objective of the design process.
Thus, the availability goal for a given system specifies certain
demands on reliability in terms of MTFB and on maintainability in
terms of MTTR. The trade-offs discussed in the next section provide
an optimum combination of these design objectives.

11.4   Trade-offs among Design Objectives
in Relation to Maintenance

Trade-off techniques are analytical processes whereby a complex
design problem, involving the selection of one of several possible
design variants, is broken down into a number of smaller problems.
Each problem is studied in the light of all the system parameters,
such as reliability, availability, safety, production, and schedule.
Finally, an optimum design solution is established. The overall objective is a life-cycle program considering various other design requirements, or data sources, as shown in Figure 11.2, including reliability
and maintainability data.

There is a definite relationship between availability, reliability, and
maintainability. Reliability and maintainability contribute in varying
proportions to provide certain levels of availability. It is evident that
if a certain degree of system availability cannot be economically
achieved through emphasis on reliability, then it can be achieved only
by incorporating better maintainability in the design.

Maintainability is the most significant factor in the eventual solution of the optimum readiness of system. Efforts to achieve system
readiness just by means of greater design reliability have often
proved to be economically infeasible. The alternative in such situations is a greatly increased emphasis on maintainability. All combinations of maintainability design features, together with the cost and

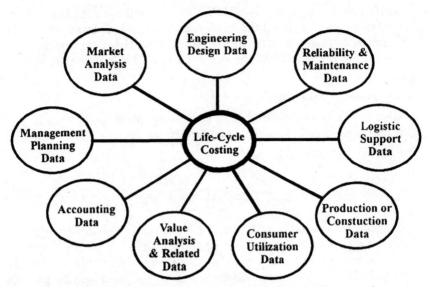

Figure 11.2 Data sources for life-cycle cost analysis. [Blanchard 95].

associated MTTR of each, must be considered to find the combinations that best meet the maintainability requirements of the system.

Following is the priority order for design objectives while carrying out trade-off studies [Anderson 90]:

1st Order   System effectiveness

2nd Order   Operational availability

3rd Order   Reliability, maintainability, support

4th Order   Parameters for reliability, maintainability, support

11.5   Rehabilitation

As opposed to the aerospace industry and other machine-related activities, civil-engineering infrastructure usually incorporates rehabilitation as a major element of the IMS. The dictionary definition of *rehabilitation* is "the action of restoring something to a former condition or status." Maintenance, on the other hand, is seen as the continuous retention of something "in an existing state," as discussed previously. These dictionary definitions imply a difference in timing of the two types of actions. It is obvious that the dividing line between the two will always remain hazy, since it is very difficult to draw a line between continuous retention and stepwise restoration. The dividing line between maintenance and rehabilitation, in practice, is often policy- and rule-dependent, and the factors affecting the dividing line can relate to the scale of work and the amount of money put into a particular activity. For example, the first 20 years of federal funding of the U.S. interstate highway program did not include maintenance, but did include rehabilitation. There is considerable evidence that under these policies some agencies withdrew support for maintenance, for lack of funds, and when the facility deteriorated enough to be defined as functionally deficient by the Federal Highway Administration rules, rehabilitation funding was provided to bring it back to current standards. The implication of "functionally deficient" in this case meant that the road or bridge needed to be upgraded so as to adhere to current standards.

Rehabilitation often seems to be related to a change in function of a facility. An old airport or airbase may be servicing only small aircraft or light loadings, but it could be rehabilitated and put into service as a heavy-duty air-carrier airport or bomber air base. In some infrastructure, rehabilitation is defined as "the process of undertaking remedial measures to solve or correct inadequacies resulting from load or natural (environmental, seismic, or hydrologic) causes." Rehabilitation, therefore, often involves changes in engineering properties, modernization, and, in many cases, changes of size, scope, function, or geometry.

Some authors point out that rehabilitation occurs when repairs and maintenance will no longer solve the problems observed. This is particularly true for a facility involving mechanical and electrical equipment, such as waste-water and water plants. In effect, continuous maintenance is unable to "retain the original state," and "restoration" to this state (or rehabilitation) is required. This also implies a difference in the size or scale of the operation. Even well-maintained infrastructure will continue to deteriorate in its functional performance to the point where it eventually requires rehabilitation or restoration.

Two main themes are apparent in practice. First, rehabilitation is generally seen as involving a larger scale of work than maintenance. Second, it involves modification or upgrading. To clarify understanding, the following concise definition for rehabilitation is offered: Rehabilitation is that set of activities involved in restoring a component system or facility to a modified or changed state.

Under this definition, filling potholes, sealing cracks, painting walls, and replacing light bulbs all remain maintenance. In fact, any actions that are directed at fixing or returning something to its *original* state should be termed maintenance under this definition. Such a definition fits well with the aircraft industry, where the term rehabilitation is not used (i.e., a 747 aircraft is maintained as a 747). It also fits one of the U.S. Federal Highway Administration's definitions of rehabilitation, in which funding is often only made available when a road or bridge is upgraded, either structurally or functionally. It thus follows that rehabilitation should occur much more often in highways, bridges, and other infrastructure that are being continually upgraded and modified, but not, for example, in the aircraft industry.

The foregoing definition highlights the consequences for infrastructure management: We cannot simply maintain facilities in a single state—we also have to continually upgrade, improve, and modernize them, either to just meet existing demand or to service continually changing demand. Some facilities, therefore, evolve during their overall lifetimes to very different forms or structures. 747s do not slowly evolve over time however into 767s! A stepwise evolution brought about by rehabilitation, in addition to the intermediate periods of continual maintenance on the maintainable parts of facilities, therefore leads to complex and interesting challenges for infrastructure managers.

11.6   Reliability-centered Maintenance

Reliability is usually defined as "the probability that a component or system will satisfactorily perform its specified function for the specified or required period of time under given or predicted operating conditions." Satisfactory performance over time is a probabilistic situation related to variations in material properties, physical environment, and load.

The reliability of a complex infrastructure facility depends on the reliability of individual components. Satisfactory system performance is achieved when all or most of the elements are performing satisfactorily. Even a slight reduction in component reliability may drastically reduce the overall system reliability.

Data regarding failure rate versus time are usually the basis of reliability predictions for infrastructure components. A system may experience different failure rates, depending upon where it is in its life cycle. It is important to know the details of the failure mechanisms and their causes, so that proper design, maintenance, or operation actions can be taken to achieve the specified reliability.

Six generic trends of failure probability against age, for a variety of electrical and mechanical items, have been reported by Moubray [Moubray 92]. These are shown in Figure 11.3. Type A is the well-known "bathtub curve." It begins with a high incidence of failure, known as infant mortality or burn-in, followed by a relatively constant failure rate, and ends in a wear-out zone. The other trends are variations on this basic theme. Generally, it has been proven that a constant failure probability, indicating that something is just as likely to fail at one time as any other (as shown by pattern E), is more applicable to electrical systems and some mechanical systems.

Probability of Failure

Time

Figure 11.3 Six patterns or types of failures. [Moubray 92].

Structural components, and therefore most public infrastructure, generally experience low or insignificant failure rates until near the end of their "structural life," at which time failure rates begin to climb (as shown by pattern B). Age is therefore observed as the major determinant of the failure probability of an infrastructure system.

11.6.1 Reliability in design as related to maintenance

Reliability is a design attribute. It means that the system's reliability is established by how well or poorly the design process is accomplished. The design, or more broadly the system definition, which also encompasses how the system must be operated and maintained, is the sole determinant in setting the inherent reliability that can be achieved. Good, sound maintenance procedures are vital to attain "potential" reliability. Maintainability is "built in," but not necessarily reliability—especially in a complex system.

The concept of reliability is a part of some contemporary infrastructure design practices. A reliability factor was introduced in the 1986 version of the AASHTO pavement design guide [AASHTO 86]. The use of safety factors in most structural design codes is an indirect way to achieve reliability in design. However, these factors of safety are basically subjective, and function as human-ignorance factors for complex design problems.

Reliability engineering advocates a comprehensive review of system specifications to assure that all objectives and supporting requirements are properly included. The design process should recognize the importance of a proper life-cycle maintenance program in attaining the inherent reliability of the system. Unfortunately, this aspect of the design process is often relegated to a secondary priority.

The application of the reliability concept is very useful in infrastructure management, where a significant relationship exists between age and performance. Although such relationships are subject to variability, the benefits in developing reasonable performance prediction models cannot be overestimated. These models tell us about the operational reliability of certain infrastructure in certain conditions. This knowledge can help in developing different construction and maintenance alternatives and allowing the selection of the most reasonable combination of initial design and future maintenance strategies.

Once the infrastructure is constructed, the subsequent management objective is to use the designed-in reliability and maintainability levels to retain the functional performance levels through the life of the infrastructure. Reliability-centered maintenance (RCM) offers a solid method to achieve this objective.

In general, RCM methodology is used to set up a detailed maintenance plan for a particular aircraft, power station, water-treatment plant, and so forth, in a rational and efficient manner. Maintenance may then continue according to this plan either until a modification takes place or throughout the operating life of the system.

The primary objective of RCM is to prioritize maintenance activities according to both the risk and consequence of failure. One of the first and most fundamental steps in setting up a detailed maintenance schedule using the RCM method is to do a failure mode and effects analysis (FMEA). This forces a breakdown of the facility into systems and components to the desired level, which is followed by a systematic assessment of the different modes of failure for each component, the probability of those failures, and the consequences. Once this has been accomplished, various maintenance actions can be assigned that will minimize the risk of serious consequences while balancing the economic and practical viability. In rare cases, the process forces redesign if fatal flaws are found where it would be impossible to reduce risks to acceptable levels. This sort of problem is often solved by designing-in redundant or backup systems such that failure can occur in the primary system, be detected, and corrective maintenance performed without necessarily affecting the overall performance.

It is not within the scope of this book to exhaustively discuss the intricacies of RCM, but it is felt to have considerable potential for application in infrastructure management.

11.7 Maintenance Management

Maintenance management is an important area of concern in overall IMS. In the broad, generic sense, maintenance management means ensuring that proper maintenance is applied at the proper time in the life cycle of the infrastructure. Over the past four decades, however, the term has taken on a more specific meaning and is usually associated with a maintenance management system (MMS). Even the term *maintenance management system* has a large number of meanings. Generally, it is similar in concept to a production control process in manufacturing, and involves setting up a database to record data, particularly cost and volumes of work, related to various maintenance activities. This was done, for example, in the 1960s and 1970s in highway departments across the United States and in other countries [Jorgensen 72, Buttler 75]. Such MMS involved recording costs, volumes of work, labor hours, and related topics associated with the day-to-day operations of highway networks.

MMSs have preceded IMSs by a number of years. In particular, the early IMSs were primarily pavement management systems (PMS) and bridge management systems (BMS). MMSs, however, were not as successful as PMSs. Although they required data collection, inventory information, and condition assessment, inadequate attention was given to inventory location and identification detail. As a result, when attempts were made to use the data to more specifically define, for example, the cost of maintenance of a particular road section, it was found impossible to separate the cost of pavement maintenance from mowing, ditch maintenance, and other maintenance activities within a linear link of highway. It was impossible to obtain data specific enough for any given length of highway—or for the pavement alone— to develop maintenance cost models for use in life-cycle costing.

Although a great deal of attention was given to propagating the systems across the United States and in many other countries, not enough attention was given to training and actual implementation into the daily work of the DOTs involved. As a result, both the inputs and outputs of most of the systems were far too broad to be of true value in an infrastructure management system.

The authors of this book have examined ten-year-old maintenance management systems in which the data were never used during that period. These MMSs were uncovered while trying to get maintenance cost information for use in life-cycle cost analysis for PMS and BMS. It was found that, in many cases, all maintenance costs for a particular mile of highway from fence line to fence line were accumulated in the same file, but there was no way to segregate specific costs of patching and crack filling, from mowing and striping, for example.

And while developing bridge management systems, the authors found that, in most cases, all costs on a short bridge were lumped together over a given period of time. It was impossible to separate data on patching versus cleaning scuppers versus rail repair.

In defining an IMS, a maintenance management subsystem (MMSS) can be developed and linked to the database of the basic IMS. In those types of infrastructure where regular operations (that is water treatment plants, sewage treatment plants, etc.) are a major element, it will be an operations and maintenance management subsystem (OMMSS). In roads and bridges, operations usually take on a more passive role, except in situations requiring significant traffic control, for example.

Maintenance-related functions involve condition assessment, preventive maintenance tasks, corrective maintenance tasks, and documentation as a minimum. The condition assessment tasks and detailing of inventory are part of the broader IMS.

Figure 11.4 shows an interrelationship between condition assessment, operations, preventive maintenance, and corrective mainte-

Figure 11.4 Linkage of operations and maintenance. [Grigg 88].

nance [Grigg 88]. All are related to the system or network in terms of the inventory.

Corrective maintenance requires a decision: Is the deficiency or damage observed serious enough to warrant entering the planning-programming-budgeting (PPB) cycle, leading to a capital budget request, or is it minor enough to take care of from field-maintenance budgets? To a degree, this depends on the level of cost involved. If the problem is major, the PPB activity incorporates information about new standards and growth forecasts, which may put the need into a rehabilitation category. There will then be a linkage between maintenance, rehabilitation, and replacement. Because of the interface in the budgeting and planning process, the same staff should be involved with planning and budgeting for rehabilitation and new facilities.

Ultimately, maintenance management involves detailed activities, such as work orders, time sheets, work-record histories, scheduling, and crew assignments, that are too detailed to be adequately treated in this book. Sixty-two cities were surveyed by the Urban Institute about their practices and experience in maintenance management. The survey results indicate that the principles of good maintenance are well known, but there has been a problem with the lack of their application [Urban 84c]. The selection of strategies involves systematic identification of available options. These include the following, as well as their combinations:

- Do crisis maintenance only.
- Maintain the worst facilities first.
- Perform opportunistic maintenance when related work is scheduled.
- Use prespecified maintenance cycles.
- Repair those components with the highest risk of failure.
- Use preventive maintenance.
- Reduce the demand for wear and tear on the facility.
- Compare the economic advantages of maintenance strategies.

Note that a rigorous application of RCM methodology, where risk and consequence of failure are vitally important, would often solve most problems associated with choosing the right strategy.

11.8 Operations as a Part of Infrastructure Management

In the broadest sense of infrastructure management, it is necessary to operate as well as to maintain facilities. Infrastructure management systems as we know them do not attempt to deal directly with operations. In other words, the systems were developed to manage maintenance within the context of the facility operation. It is very important, therefore, to realize that the environment in which the IMS is used, and indeed the reason for its existence, is the operations it must support. This operating environment must obviously be considered when planning bridge or road maintenance, since traffic must be delayed, detoured, or stopped. Other types of infrastructure require even closer cooperation between the management system and the operating environment. Buildings, dams, electrical facilities, as well as airports, air bases, and chemical and petroleum manufacturing plants are some examples.

Grigg explains some of the interaction between an IMS and its operating environment and shows that the initial design of the IMS is dependent on the target performance objectives of the agency, which are, in turn, often influenced by the effects of bureaucracy [Grigg 88]. In the vast majority of cases, at least three distinct phases characterize the life of a facility: (1) planning and design; (2) construction; and (3) operation and maintenance. The latter is generally the longest and often the most important. It should always be remembered that the initial design should include a maintenance schedule, which implies that this schedule must take due cognizance of the operations and that flexibility is greatly reduced after construction.

The concepts outlined by Grigg and other authors point out the value of setting goals and objectives and defining methods and rules that will motivate the staff workers dealing directly with the IMS. This is a critical part of an effective infrastructure management process.

The whole purpose of an operations management activity is to provide the maximum productivity within the organization. This leads to concepts of performance measurement, effectiveness, and efficiency. It is helpful to think of the output of an organization in terms of production. For those types of infrastructure that have a major operations component, it is important to measure effectiveness in terms of the actual production compared to the theoretical production.

11.8.1 Quality control and quality assurance

Quality control (QC) is a vital part of operations as well as all other aspects of infrastructure management, particularly construction. Quality control has become increasingly important as we learn that high-quality products are not only the best way to compete, they are essential [ASCE 88]. We have also learned that the best way to have high quality is to do things right the first time. There is a lot of activity in infrastructure today associated with quality control, quality assurance (QA), and continuous quality improvement or total quality management (TQM).

11.8.1.1 Definitions. The following are definitions of quality control and quality assurance:

Quality Control (QC) stands for operational techniques and activities aimed both at monitoring a process and at eliminating causes of unsatisfactory performance. QC refers to technical detail and is the method manufacturers and builders use to ensure that their materials are up to standard and that products are fabricated or built correctly.

Quality Assurance (QA) is defined as "all planned actions necessary to provide adequate confidence that a product or service will satisfy given requirements for quality." Quality assurance is the process used to ensure that quality controls and service aspects are carried out, and to monitor and record the appropriate verifications applicable to activities during each phase of a project.

Quality assurance is a key consideration in end-result specifications (ERS), based on statistical concepts. The growing use of performance-related criteria in ERS should evolve to long-term performance requirements. This would be a move in the direction of European contracting procedures.

The quality of materials incorporated and work performed directly influences the service life of an infrastructure, maintenance costs, level of service, and user costs. Quality can be defined as "the characteristics of a product (materials and methods of fabrication, for instance) that provide a level of functional (i.e., serviceability) and structural performance and design life." QC ensures that the specified "ingredients" are combined and placed so that the product will have the desired level of performance. QA is all of the activities necessary to verify, audit, and evaluate quality. With the move to ERS, the contractor/subcontractors/suppliers take on QC responsibility (such as process control), and the agency/consultant uses QA for assessing acceptance.

11.8.1.2 Scope of quality control. The scope of QC in an ERS framework, from a contractor or supplier's viewpoint, involves the following:

- QC is the method used to ensure the product/work will be accepted.

- Contractors or suppliers impose a QC system of inspection, sampling, and testing (process control) on themselves.

- Without QC, a contractor or supplier does not know whether a proper level of quality is being achieved until the product/work is either accepted, price-adjusted, or rejected by the agency.

- QC may not be specified by the agency, but is essential.

11.8.1.3 Scope of quality assurance. The scope of QA from an agency (owner) viewpoint can be summarized as follows:

- What do we want? (planning and design)
- How do we order it? (construction plans and specifications)
- Did we get what we ordered? (inspection, sampling, and testing)
- What if we don't get what was ordered?
  1. Accept substandard (marginal or borderline).
  2. Do not accept (repair or reject).
  3. Assess price adjustment (penalty).
- What if we get "better" than ordered? (bonus?)

11.8.2 Total quality management (TQM)

With a move toward ERS or warranty specifications, the need for improving the quality of work increases. Improved quality will have a direct impact on bonus/penalty provisions, and contractors are accordingly seeking assistance through the new philosophy of TQM. Partnering, an integral part of TQM, is rapidly becoming a project strategy for agency-contractor commitment and communications.

TQM incorporates all of the activities associated with the continuous improvements of quality and productivity (service), with an emphasis on customer-supplier relationships, employee involvement in decision making, teamwork, rigorous analysis of work as a process, statistical quality control, and managerial focus on leading. This is essentially the definition of TQM that has been suggested for the engineering and construction industry [FHWA 93].

11.8.2.1 Principles of TQM. According to Deming, the following obligations must be fulfilled by executive management to improve quality, productivity, and competitive position [Deming 86]:

1. Create constancy of purpose toward improvement of service and product to be competitive, stay in business, and provide jobs.
2. Adopt the new philosophy. Management must awaken to the challenge, learn their responsibilities, and accept leadership for change.
3. Create sole dependence on inspection to achieve quality. Eliminate the need for inspection on a mass basis by building quality into the product in the first place.

4. End the practice of awarding business on the basis of a price tag. Instead, minimize total cost. Move toward a single supplier for any one item, on a long-term relationship of loyalty and trust.

5. Improve constantly and forever the system of service and production, to improve quality and productivity and thus constantly decrease costs.

6. Institute training on the job.

7. Institute leadership (management and production workers) to help people and machines do a better job. Leadership of management is in need of overhaul, as well as leadership of production workers.

8. Drive out fear, so that everyone may work effectively for the organization.

9. Break down barriers between departments, so that people work as a team.

10. Eliminate slogans, exhortations, and targets for the work force, asking for zero defects and new levels of productivity.

11. Eliminate work quotes on the production floor; eliminate management by objective and by numbers (numerical goals). Substitute leadership.

12. Remove barriers that rob hourly workers, management, and engineering staff of their right to pride of workmanship.

13. Institute a vigorous program of education and self-improvement. Start with executive managers.

14. Put everybody in the organization to work to accomplish the transformation.

The following important concepts are involved in the TQM philosophy:

Quality Strategy—The most important aspect of TQM is the commitment of the decision-makers and executive managers.

Customer Orientation—This is the core of TQM. Customer means both the internal (within organization) as well as the external (end-users and public) customers. (Public agencies have customers, too.)

Employee Involvement—Employee involvement in organizational activities and other steps of the process is equally important. People conceive innovations. It means the involvement of all employees in activities relating to quality, including analysis of a particular problem and finding solutions.

Process-centered Approach—TQM shifts the emphasis from the earlier concepts of the quality of product or quality of service to the process itself. If the process is of high quality, then the product is bound to be of high quality.

Problem-solving Tools—The basic management tools also apply for TQM. These basic tools include statistical techniques, check sheets, histograms, cause-and-effect diagrams, design of experiments, multivariate analysis, stratification, control charts, design review, failure analysis, and analysis of alternative solution strategies.

Continuous Improvement—Every organization requires improvement, which is the most important aspect in implementing TQM.

Benchmarking—It is a continuous process of measuring one's own products and services, business practices, and related activities against the leading competitor.

People Orientation—It is the mutual trust of the employee and the management in good and bad times that leads to a motivated and dedicated team.

Life-cycle Maintenance Plan—Life-cycle maintenance plans reduce breakdowns, maximize machine operating efficiency, increase life and encourage maintenance-free designs, and help to build up a trained and motivated team.

11.8.2.2 TQM's implementation and costs. Without an adequate quality-assurance system, fluctuating levels of quality may result in extra production costs. The installation of a quality system like ISO 9000-9004 standards may stabilize the quality at a predetermined level. However, TQM provides for constant quality improvement through a strong customer focus, as opposed to mere conformity to specifications.

It is expected that the implementation of a TQM will result in increased profit and less employee turnover while reducing costs, loss of breakdown time, rejections, and complaints. Figure 11.5 shows the concept of reduced quality costs for process checking versus product checking [Hayden 89].

Considering infrastructure management as a process, and implementing TQM concepts, provide numerous benefits, including the possibility of aggressive commitment from executive decision-makers and the participation of the public. The City of Louisville airport expansion project is an excellent example of active participation by customers and the public in the planning and implementation stages [Michael 96].

11.8.3 Role of ISO 9000 standards

A new series of quality standards called ISO 9000 has been created by the International Organization for Standardization (ISO). Its adoption by the European Economic Community has dramatically increased the use of quality-assurance programs. The ISO 9000

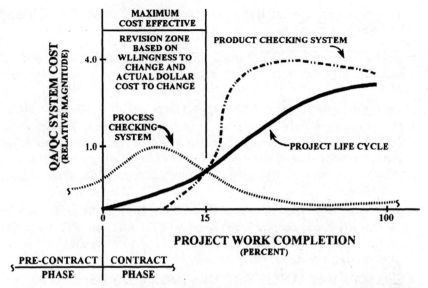

Figure 11.5 Process checking versus product checking quality systems (after [Hayden 89]).

Series of Quality Standards is now the worldwide accepted quality-assurance standard, but it is still in the evolutionary stage. In essence, it is a steppingstone toward a full TQM process. Future revisions and supplements will be published to address issues that are raised as more businesses learn and use these standards. There are several categories, including ISO 9001 to 9004. The appropriate quality-assurance category should be selected considering the following six factors:

1. Design-process complexity
2. Design maturity
3. Production-process complexity
4. Product/service characteristics
5. Product/service safety
6. Economics

While many companies in the U.S. are becoming ISO 9000–registered, the quality system requirements specified in ANSI/ASQC Standards Q91, Q92, Q93, and Q94 are complementary to the technical specified requirements of a product and/or service. These standards are technically equivalent to the international standards ISO 9001, 9002, 9003, and 9004, respectively [ASQC 87].

Following is a brief list of the applicable areas for ISO and ANSI/ASQC standards.

- **ISO 9000-1987 (ANSI/ASQC Q90-1987),** *Quality Management and Quality Assurance Standards—Guidelines for Selection and Use.*

- **ISO 9001-1987 (ANSI/ASQC Q91-1987),** *Quality Systems—Model for Quality Assurance in Design / Development, Production, Installation, and Servicing.* For use when conformance to specified requirements is to be assured by the supplier during several stages, which may include design/development, production, installation, and service.

- **ISO 9002-1987 (ANSI/ASQC Q92-1987),** *Quality Systems—Model for Quality Assurance in Production and Installation.* For use when conformance to specified requirements is to be assured by the supplier during production and installation.

- **ISO 9003-1987 (ANSI/ASQC Q93-1987),** *Quality Systems—Model for Quality Assurance in Final Inspection and Test.* For use when conformance to specified requirements is to be assured by the supplier solely at final inspection and test.

- **ISO 9004-1987 (ANSI/ASQC Q94-1987),** *Quality Management and Quality Systems Elements—Guidelines.* For design of an effective quality-management system to satisfy customer needs and expectations while serving to protect the company's interests. A well-structured quality system is a valuable management resource in the optimization and control of quality in relation to risk, cost, and benefit considerations.

Among several national efforts in the U.S. to promote quality, AASHTO has begun to develop a QA/QC guide specification for the highway industry. This guide follows the efforts of numerous states that have developed and used quality-assurance specifications, some since the mid-1960s [FHWA 93]. ISO 9000 reinforces the importance of employee and management participation in the quality-production process. Integrating the TQM principles of employee participation and continuous process improvement with the ISO 9000-9004 standards can give any organization better tools.

References

[AASHTO 86]   "AASHTO Guide for Design of Pavement Structures," American Association of State Highway and Transportation Officials, Washington, D.C., 1986.
[Anderson 90]   R. T. Anderson and L. Neri, *Reliability Centered Maintenance: Management and Engineering Methods,* Elsevier Applied Science, New York, 1990.

[ASCE 88]   American Society of Civil Engineers, *Quality in the Constructed Project,* manual of professional practice, Vol. 1, 1988.

[ASQC 87]   American Society for Quality Control, *American National Standard, ANSI/ASQC 094-1987, Quality Management and Quality System Elements— Guidelines,* Milwaukee, Wis., 1987.

[Blanchard 95]   B. S. Blanchard, D. Verma, and E. L. Peterson, *Maintainability: A Key to Effective Serviceability and Maintenance Management,* John Wiley and Sons, New York, 1995.

[Buttler 75]   B. C. Buttler and L. G. Byrd, "Maintenance Management," *The Handbook of Highway Engineering,* Section 25, Van Nostrand Reinhold, New York, 1975.

[Deming 86]   W. E. Deming, *Out of the Crisis,* Massachusetts Institute of Technology, Cambridge, Mass., 1986.

[DOD 66]   "R and D Material: Maintainability Engineering," *Publication 705-1,* U.S. Department of Defense, Washington, D.C., 1966.

[FHWA 93]   Federal Highway Administration, "Construction Quality Management for Managers," *Publication No. FHWA-SA-93-071,* Demonstration Project 89, Office of Engineering and Office of Technology Applications, Washington, D.C., 1993.

[Grigg 88]   N. S. Grigg, *Infrastructure Engineering and Management,* John Wiley and Sons, New York, 1988.

[Hayden 89]   W. M. Hayden, *The Effective A/E Quality Management Program: How To Do It,* short course notebook, American Society of Civil Engineers, New York, 1989.

[Jorgensen 72]   Roy Jorgensen Associates, "Performance Budgeting System for Highway Maintenance Management," *NCHRP Report 131,* Transportation Research Board, National Research Council, Washington, D.C., 1972.

[King 86]   F. H. King, *Aviation Maintenance Management,* Southern Illinois University Press, Carbondale, Ill., 1986.

[Michael 96]   R. S. Michael, "Building Support for Airport Expansion," keynote address, 24th International Air Transportation Congress, Louisville, Ky., June, 1996.

[Moubray 92]   J. Moubray, *Reliability Centered Maintenance,* Industrial Press, 1992.

[Smith 93]   A. M. Smith, *Reliability Centered Maintenance,* McGraw-Hill, New York, 1993.

[Urban 84c]   H. P. Hatry and B. G. Steinthal, *Guides to Managing Urban Capital Series,* Vol. 4: *Guide to Selecting Maintenance Strategies for Capital Facilities,* The Urban Institute, Washington, D.C., 1984.

[Weeks 96]   K. D. Weeks and H. W. Jandl, " The Secretary of the Interior's Standards for the Treatment of Historic Properties: A Philosophical and Ethical Framework for Making Treatment Decisions," *Standards for Preservation and Rehabilitation, ASTM STP 1258,* American Society for Testing and Materials, Philadelphia, Pa., 1996, pp. 7–22.

12

Dealing with New
or Alternate Concepts

12.1 Introduction

For many years, civil engineering has dealt primarily with groups of standard materials, including conventional steel, Portland cement concrete, asphalt concrete, vitrified clay (brick and pipe), glass, wood, and naturally-occurring aggregates. Other materials are also used, and occasionally new materials are introduced. Traditional design, maintenance, and construction procedures have all been built around these conventional materials and their combinations, such as reinforced concrete and jointed concrete pavements. Historically, structural design has been associated with the design of conventional structural steel and conventionally reinforced concrete. Several generations of engineers have been trained using these materials, and textbooks have been written covering the concepts related to such materials.

Design guides and design manuals are also formulated around these conventional materials. Examples include the Portland Cement Association's pavement design manual [PCA 84], the various design manuals produced by the Asphalt Institute [AI 87, AI 91], the American Institute of Steel Construction design handbook [AISC 94], and the American Concrete Institute handbooks of various types [ACI 72, ACI 84, ACI 92].

Of course, a variety of new materials have been tried in the last century. However, the frequency with which they have been used successfully, and the difficulty of adapting new materials, has often hampered new developments.

A good example is the California Bearing Ratio (CBR) design method for asphalt pavements, which was widely used by the U.S. Army Corps of Engineers and the Asphalt Institute prior to 1965, and is still being used around the world [USACE 58, USACE 61]. The CBR method is an empirical method based on the use of conventional crushed-stone materials covered by a relatively thin layer of conventional asphalt concrete. In 1962, the AASHO Road Test was completed [HRB 62]. This study showed that the value of a material for carrying repeated traffic loads depended on the quality of the material itself, not just the thickness. The CBR method does not take this concept into account, and has now been generally replaced in use. In a similar fashion, various aspects of Portland cement concrete pavement design were shown by the AASHO Road Test to need adjustment and were modified in the AASHTO Pavement Design Guides, which are still in use [AASHTO 93].

Similar examples can be cited for the building industry, as empirical methods developed by Ferguson et al. [Ferguson 88] improved on the empirical knowledge about the use of steel reinforcement in the design of reinforced concrete structures and building elements. This development has continued with the advent of higher-strength concrete, finer-grind Portland cements, and high-tensile-strength steel.

Many new and innovative materials, like fiber-reinforced plastics (FRP), are not readily accepted by the specification and design communities. The most significant reason is the time lag between the development and evidence of long-term performance. Another reason is that the use of existing design standards with new material technology may not be the best way to implement the technology; it may be more costly and even fail. These points are obvious if one considers bridge technology. Bridges have always been a permanent record of mankind's achievements in the use of materials [Bridge 95]. After each new material was developed, it took 30 to 40 years for it to gain the acceptance and confidence of engineers before they would use it for a long-span bridge design. For example, the first iron bridge was built in 1779, very soon after the invention of wrought iron, but the material was used in the first world-record span in 1816.

As illustrated in Figure 12.1, increases in bridge span have occurred in three jumps: one in middle 1800s when wrought iron was sufficiently developed; one in the late 1800s when reliable steel was available; and the largest of all in 1930 when high-strength steel wire was used in a major suspension bridge. The trend indicates that the maximum span has risen by a factor of 3.5 in each 100 years since 1770 [Bridge 95].

Steel is always one of the first choices for structures of the future because of improvements in welding and joint technology and higher yield strength. Concrete technology has also improved rapidly, resulting in very high strength in a shorter time. However, the advanced composites and FRP materials provide new concepts in structural forms, resistance against corrosion, and higher strength-to-weight ratio, as shown in Figure 12.2 [after Jacobs 94].

Figure 12.1 Historical trend in the world's longest bridge span construction. [Bridge 95].

Figure 12.2 Historical development of engineering materials. (after [Jacobs 94]).

12.2 Examples of New Material Usage

There are numerous examples of attempts to use new materials in conventional infrastructure—some successful, some not. Examples of successes include the use of high-strength steel and high-strength concrete, and the introduction of continuous steel reinforcement in pavements with the advent of continuously reinforced concrete pavements (CRCP) in the 1950s. Even this last example of CRCP, although widely used, shows the difficulty in introducing new concepts as the result of a design-only process.

12.2.1 CRCP pavements

CRCP was first introduced in reasonable quantities in the state of Texas in the 1950s [McCullough 60, 68]. From early success it appeared to be a panacea because of its very smooth riding quality and its apparent ease of construction without joints. However, problems soon developed in the as-constructed product.

The original design was based upon strength characteristics of the steel and concrete in which the steel was placed at the mid-depth of the concrete in sufficient quantity to hold cracks tightly together, but the concrete itself cracked to relieve concentrated tensile stresses. This part of the design worked well; however, the design did not account for the significant increase in deflection that occurred with the thinner reinforced slabs with no joints. The resulting deflection increase later caused failure in the subbase and subgrade under the pavement within a period as short as two to five years. This loss of support then resulted in even higher deflections in the pavement and subsequent punch-outs and failure. The Texas Department of Transportation (Texas DOT) immediately spotted these problems, subsequently modified their designs, and have continued to build the pavements successfully. The revised design produces pavements in general 51 to 102 mm (2 to 4 in) thicker than the original 203-mm (8-inch) thick design. This improvement was the result of the interaction of design, construction, maintenance, and data feedback.

Another problem associated with CRCP was the need for different construction techniques. When CRCP was first developed, most highway contractors had conventional form-type paving equipment. However, CRCP lends itself to formless paving because of the continuous nature of the steel. When pavement contractors learned how to build the new pavement, a rapid reduction in construction costs resulted. Unfortunately, such success did not occur on major projects in Florida and Indiana, where construction, inspection, and supervision was reportedly inadequate. These projects resulted in devastating early failure, which caused the FHWA to cancel recommendations for constructing this type of pavement. As of 1996, this recommended support is still suspended. If an effective infrastructure management system had been in effect as this new concept was being developed, it may well have provided codified technology for dealing with not only the design concept changes but also the construction and maintenance requirement changes.

12.2.2   CORTEN steel building construction

Some new materials may be valuable in the proper setting, but misused. One such example involves introducing a new building material on the campus of the University of Texas at Austin. The concept of CORTEN steel [Bradford 93] is that it is "designed to oxidize or rust." Such oxidation is supposed to occur quickly and uniformly to produce a thin, hard, impervious layer, which presumably prevents further rusting and deterioration of the steel. In 1970, an architect accepted the recommendation of steel manufacturers to use some of the new steel as a facing material on the Communication Building at the University of Texas in Austin. This material rusted as planned and formed a reddish-brown color. However, it did not stop rusting as predicted, and subsequently rusted completely through within a period of five to six years. The result was that many sections had to be replaced, and eventually a complete new surface cladding had to be installed on the building at tremendous cost. This material has been used successfully as a structural material, but was not acceptable in thin layers.

12.2.3   Chemcrete asphalt concrete mixes

Chemcrete is another example of premature acceptance of a new concept without adequate evaluation. In the late 1970s, a major company undertook the development of a "new" material called Chemcrete. The material was purported to be a panacea for asphalt concrete construction. It produced very stiff pavements, which its proponents labeled "a stronger pavement." Unfortunately those involved did not adequately recognize that asphalt concrete mixes must fulfill several purposes to perform successfully. These include not only stability, but also strength, flexibility, and durability. The company involved enlisted several well-known professionals in the pavement field to help promote their product. The laboratory tests and the field trials, however, were observed for only very short periods of time, so the durability aspect of the material was not properly evaluated. Subsequent pavements using Chemcrete cracked badly and rapidly, and resulted in extremely early failure. The product ceased to be marketed a few years later. Proper application of an infrastructure management system, or in this case a pavement management system, would likely have shown the deficiencies at a very early date during the test period.

12.2.4   The use of geotextiles

Another example of materials use that has both positive and negative aspects involves fabric or geotextiles in pavement construction.

Since 1960, various manufacturers of nonwoven fabrics (mostly using polypropylene fibers) have been seeking to broaden the markets for their products. Pavements appeared an attractive target, and they began vigorous efforts to market their product for use in pavement rehabilitation. These materials have been marketed as a cure-all that will waterproof (because of the application of a tack coat on the existing, old pavement) and prevent reflection cracking in overlaid asphalt concrete pavements. Television and trade-magazine advertisements have appeared depicting the apparent speed and quality with which various city engineers were supposed to be able to repair their streets using these products. The ads never mention that the cracks do not disappear, but are only delayed in returning by one or two years. They also do not refer to the fact that the use of a fabric is the equivalent in cost to one or more additional inches of asphaltic concrete, which would show better pavement performance.

On the other hand, there is a great deal of evidence of successful and effective use of fabric as a construction technique when placed as a separating medium on wet or soft subgrades. The use of the fabric or geotextile provides a platform for placing granular material that will not punch into the soft subgrade, and which saves considerable money, time, and materials over waiting for drier conditions or compacting the aggregate into the soft subgrade. This use of fabrics can be valuable and meets the test of economic expediency. Fabrics are also successfully used as drainage layers and for confining drainable trenches in some cases, although clay particles can tend to clog the fabric and prevent the desired drainage.

12.2.5 Applications of polymers in construction

12.2.5.1 Polymers.
The benefits of polymer material technology are evident in many areas of life, including sports gear, auto parts and autobody repairs, buttons, aerospace industry, bikes, musical instruments, and home furnishings. Most items are made of cast polymer or fiber-reinforced composites.

12.2.5.2 Cast-polymer products.
The use of polymers and particulate composites is not new in construction applications. Cast-polymer products have been used for everything from home furnishings, such as kitchen countertops and bathrooms, to rapid repair of bomb damage [Dawson 95].

Cast polymer is a nonreinforced particulate composite that combines polymers, fillers, and additives to meet specific application requirements. Fillers account for up to 90 percent of the material volume. After the material is mixed, it polymerizes and becomes rigid [Dawson 95].

Methyl methacrylate acrylic (MMA) and unsaturated polyesters resins (UPR), both orthophthalic and isophthalic, dominate the gel-coated cultured marble and solid surfaces markets, which include home furnishings for kitchen, bathrooms, and whirlpool spas. Solid surface densified and nonporous products made from high-quality resin can be cut, sanded, and seamlessly joined. Architects and specifiers can take full advantage of these properties, as well as resistance to corrosion and chemicals, for creative architectural and interior designs.

Orthopolyester and polyurethane are commonly used for decorative castings and flexible moldings. Polyester, epoxy, and vinyl ester are the predominant matrix resins in precast polymer concrete for products such as utility vault doors and electrolytic mining cells. They also are used in field applications, such as the polymer concrete composite overlay used for the Golden Gate Bridge approach [Dawson 95]. A polymer concrete overlay can be opened to traffic within a few hours compared to the 14 to 28 days curing time required for conventional Portland cement concrete.

12.2.5.3 Polymer modified asphalt.   Asphalt is used with sand and selected sizes of aggregates as hot-mix, emulsion and cold mix, and seal coats for paving parking lots, streets, highways, and airport pavements. On heavily trafficked roads, rutting has become a major distress and reason for poor performance. Polymer-modified asphalt is becoming increasingly popular to improve the thermo-viscoelastic behavior and in-service performance of asphalt pavements. Mixes paved with asphalt or polymer-modified asphalt are examples of particulate composites, and their stress-strain behavior can be analyzed using the same micromechanical analysis used for advanced polymeric-matrix fiber-reinforced composites [Plaxico 95, Plaxico 96].

12.2.6 Fiber-reinforced plastics (FRP) for structural applications

12.2.6.1 Overview.   For over three decades, polymeric fiber-reinforced plastics (FRP), broadly known as *advanced composites,* have been used for grates, steps, ladders, pipes and tanks, and numerous other applications in petrochemical plants and chemical industries where resistance to corrosive chemicals is a primary concern as compared to structural performance. In these applications, composites outperform steel, concrete, and wood. The aerospace industry has made use of these advanced composites in nonstructural components because of their light weight, higher strength-to-weight ratio, and superior resistance to temperature changes and fire [Technology 95].

The application of polymeric advanced-composite materials has been rather slow in civil-engineering structural market and construction industries. IMS can show owners, operators, and designers the importance of low maintenance, longer life, and improved structural performance offered by advanced high-performance construction materials. The application of advanced composites is being sought in all areas of construction, particularly for waterfront and off-shore structures and seismic retrofitting of bridges. Another application area is the rehabilitation of reinforced concrete structures, where the major problem is reduced life due to the corrosion of steel reinforcement in a harsh environment. Use of FRP products could significantly increase the life and performance of these structures. The light weight of FRP products makes handling easy and reduces assembly costs. The FRP materials made from polymeric resin matrix and fibers are becoming popular, as the trend toward achieving low-cost manufacturing methods increases.

12.2.6.2 Selection of FRP material and application process. The proper use of FRP material and application process is the key to successful and longer-lasting applications in structural strengthening and retrofit of infrastructure assets. FRP composites are anisotropic in nature, and this should be used to advantage in innovative designs. Three primary limitations of FRP are long-term loss of strength due to creep effects; debonding and fiber material degradation due to the likelihood of water absorption by the resin matrix; and durability problems in an alkaline environment. These should be appropriately considered in design applications. The following examples illustrate innovative applications of FRP materials.

Naval FRP Piers The U.S. Naval Facilities Engineering Service Center (NFESC) installed an all-composite catwalk at its Advanced Water Technology Test Site (AWTTS) at the Port Hueneme, California, site in May 1995 to test composites in marine structures. The total length of the test pier is 48 m (160 ft), consisting of several spans. The composite portion of the new pier consists of two bays: a fiberglass composite deck and a concrete deck prestressed with carbon-fiber/epoxy rods. The decks are the result of a cooperative Construction Productivity Advancement Research (CPAR) project among the Composites Institute, the Army Corps of Engineers CERL, and the South Dakota School of Mines and Technology [Technology 95, Iyer 95]. The deck was designed to carry pedestrian loads, withstand seawater exposure and even total submersion during extreme high tides, and be subjected to cyclic loading from wave action during tide charges.

Seismic Retrofit of Bridge Columns Seible and Innamorato [Seible 95] identified the possible failure modes of concrete bridge columns based on a study sponsored by the California Department of Transportation (Caltrans). For these cases, the three primary failure modes are: (1) column shear failure; (2) concrete confinement failure of the flexural plastic hinge region; and (3) debonding of lap splices in the column reinforcement.

A reliable retrofit measure should consider all three possible failure modes. Such retrofit methods include steel column casing to jacket bridge columns. For conventional columns with a uniform circular cross section, the steel jacket can be fabricated in two halves, bolted together, and filled with grout. For nonuniform cross sections or other shapes, steel casings are difficult to fabricate. For strengthening and rehabilitation of bridge columns, a composite wrapping and winding system is an efficient alternative, from a functional viewpoint, to provide hoop constraint. Seible and Innamorato [Seible 95] outline the principles of an automated continuous carbon-fiber wrap system that addresses the transverse reinforcement deficiency by wrapping prepregnated carbon tows in the horizontal, or 90 degree, direction to the column axis to provide the required transverse confinement, clamping, and buckling restraints. This innovative composite retrofit system ensures anchorage of the wound carbon tows by the continuity of the fiber wrap for the entire column jacket, and provides close control and monitoring of lay-up thicknesses with an automated winding system.

12.2.6.3 Responsibility of specifying and approving agencies.
Development of acceptable test standards and specifications for the emerging technologies, such as FRP composites, is a major concern for the specifying and approving agencies. In particular, agencies are interested in the time-critical emergency situations that warrant structural applications in life-line infrastructure facilities, such as seismic retrofitting of bridges and repair and strengthening of structural members. Caltrans [Sultan 1995] has already initiated an active program in this area for performance testing of FRP to verify

- Strength and performance of the fiber itself
- Strength and performance of the resin
- Strength and performance of the composite
- Behavior of the system as a structural member
- Validity of the design approach

Durability of FRP composites should be checked for resistance to water, ultraviolet rays, high temperature, salt water, alkaline environment, freeze-thaw, and fire. Table 12.1 outlines a list of important issues that an approving agency should examine to qualify and imple-

TABLE 12.1   Checklist for Qualifying and Implementing FRP
Technology [Sultan 95]

1. Product documentation
2. Understanding of the process
3. Type of materials
4. Material physical properties
5. Long-term durability (chemical and physical) testing of composite
6. Quality control in manufacturing, mixing, and applying
7. Fiber content, voids, resin ratio
8. Design guideline for the specific composite
9. Safety factors
10. Damage and failure modes
11. Adequate specifications
12. Repeatability and consistency
13. Acceptable field erection methods
14. Effect of fatigue on bond behavior
15. Performance under dynamic load
16. Testing under sustained loading
17. Qualifications of suppliers and product designers
18. Cure temperature
19. Transportation and handling
20. Maintenance issues

ment any proposed FRP material-fabrication-installation process
[Sultan 95].

12.3   Handling Diminishing Resources

During the 1940s, 1950s, and 1960s, the United States and many other
areas of the world were on a veritable building binge. Notable construction included the interstate highway system, the rapid development of
Toronto, New York, Chicago, Los Angeles, and other large cities, including buildings, sewage, water supply, dams, and related infrastructures.
This construction grew out of the completion of the greatest war ever
fought, World War II. Techniques developed during the war for rapid
construction of facilities and equipment at any cost continued to be
used in the creation of the new national infrastructure. Although this
approach was successful in getting construction done, it resulted in a
massive depletion of the world's natural resources.

The construction of the U.S. Interstate highway system alone consumed a large quantity of good natural aggregates in many sections

of the United States. An outgrowth has been the creation of tighter and tighter specifications requiring higher and higher quality materials, which have become more difficult for the manufacturers and producers to provide. In some cases, this has made costs prohibitive. An infrastructure management system can properly consider both the benefits and costs of the materials, with their limitations, and provide an effective and economical way for using some of the so-called marginal materials.

Since the early 1980s, the world, and in particular North America, has become concerned with environmental issues and the reuse, reclaiming, and recycling of certain materials. This problem has two aspects: (1) the required change of material usage due to environmental issues; and (2) the potential successful use of reclaimed, recycled, or otherwise waste materials.

12.3.1 Environmental impacts

In the last few decades, much has been learned about the environmental impact of certain materials used in infrastructure development. A well-known example is the use of asbestos. For about 50 years prior to 1970, asbestos was used as an insulation material in buildings, particularly around heating ducts and heating pipes, and as a fire-resistant material in the form of sprayed-on insulation around structural steel and reinforced concrete beams. It was also used as a siding and roofing material in the form of pressed asbestos sheeting and corrugated paneling, and even as a filler material in asphalt concrete to replace certain types of fine aggregate. Environmental tests subsequently showed asbestos to be carcinogenic. Its use in construction was suspended, and efforts began to remove it from many construction locations. Two aspects of this problem could be handled by an appropriate infrastructure management system: (1) the development of new materials to replace asbestos; and (2) maintenance and removal procedures for dealing with the material in place. Unfortunately, the basic and very costly remedial approach has been to remove the material. However, a thorough evaluation of the costs and benefits might conceivably show that the benefits of removing the material are minimal, and that the damage done by disturbing the material and removing it is greater than would ever be caused by sealing it and leaving it in place.

This is an example where external factors—that is, public fear and the profit motive for certain industries—has overridden good systems analysis. This is not to say that the environmental aspects should be ignored, but merely that they should be thoroughly evaluated in a realistic sense along with the total costs and benefits.

12.3.2   Environmentally safe alternatives

The asphalt industry provides examples of existing materials that proved to be environmentally questionable, and the environmentally safe alternatives were developed in their place. Specifically, cutback asphalts, those asphalt cements that are diluted with gasoline and diesel fuel, were widely used in the United States and around the world prior to 1970 for constructing low-volume roads and for priming subgrade and base layers on high-quality construction. The volatile component of the cutback asphalt, i.e., gasoline and diesel, evaporate into the atmosphere, thus adding to the problems of air quality, particularly in urban areas. Alternate methods of changing the viscosity and transportability of asphalt are: (1) to heat the material; and (2) to dilute it with water in the form of an emulsion. Considerable efforts have been made in the past 30 years to improve the quality of emulsified asphalts to replace asphalt cutbacks.

12.4   Energy-related Issues

Energy consumption is sometimes a separate and more important issue in selecting infrastructure design, construction, and maintenance techniques than economics alone. In oil-importing countries, such as Chile, while the economics of asphalt pavements may be acceptable, the strategic issues and support of the local Portland cement industry may suggest the application of a systems concept. Even in the United States, which is relatively energy-rich, the balance of payments strongly suggests the need to examine the energy efficiency of a system as well as its overall performance and economy [Solminihac 92, Jones 93].

Infrastructure Management Systems (IMS) offer methods for managing roadway infrastructure to reduce maintenance problems, rehabilitation delays, and lane-closure times. Thus, energy can be saved through improved infrastructure management. A coordinated methodology for planning, design, construction, and maintenance of transportation infrastructure minimizes vehicle delays and congestion and saves millions of gallons of gasoline and diesel.

Technically speaking, there is no difference between an IMS with and without energy savings taken into account. The difference lies in the variables used in the systems. In general, more attention should be paid to decision variables related to current and future user cost, fuel consumption, and so forth, to calculate energy savings. Although variables related to energy saving may be difficult to quantify, they can be used indirectly in decision making. Just like other issues, energy consumption can be considered in all the activities from planning, design, and construction to maintenance.

Regarding pavements, attention should be paid to better network-level planning design, material use, construction, and maintenance to save energy over the long run. Better design increases pavement life, thus decreasing the number of rehabilitation cycles and saving fuel due to reduced delay during pavement construction. Another method of saving energy is to recycle existing material. Recycling existing material saves the energy that would be required to make new materials to replace them. It further reduces the need for energy and the costs associated with the disposal of old materials.

The Urban Roadway Management System (URMS) developed by Chen et al. at the University of Texas [Sohail 95] was one of the first comprehensive pavement-management systems for application in small to medium-size cities. It provides managers and engineers in public-works departments with a computer-based tool to assist in all their activities, including planning, design, construction, and maintenance. For agencies desiring to evaluate associated "energy costs," the URMS can be adopted to minimize energy use as well.

12.5 Considering New Methods and Materials for Infrastructure Use

How does an agency involved in the provision of infrastructure go about considering new materials, methods, and designs? No agency wants to be blind to new innovations, but neither do they want to fall prey to wild marketing claims and promises. The best approach to this problem is a formalized IMS. During the interim, however, general systems-management concepts can be of assistance. The first step would be to look at the standard or typical procedures that are currently used. The second step would be to define the alternatives that should be compared to the standard. The third step is then to list all of the criteria that must be used in making the comparison. Performance, strength, stability, and service life, of course, immediately come to mind, but in some cases the more important criteria may be maintainability, recycling potential, use of local materials, environmental impact, energy impact, and political necessity. It is important to list these factors á priori, that is, before comparisons are made.

A thorough and complete listing of factors and costs (not only economic costs) involved in any issue is critical. If a preliminary comparison is made that does not consider all pertinent factors, it is very difficult to readjust the final solution based on the addition of other factors. It is always tempting to be biased to the initial solution, or to add those factors that point to the solution you prefer if the "first solution" does not seem suitable. Such an approach is not objective, and is far less likely to give meaningful results than a truly objective approach.

A well-defined infrastructure management system will provide the listings and approaches necessary to include the majority of factors from the outset. However, no system is perfect, and the judgment of the infrastructure management team is critical in setting up the problem. It will be preferable, and a better solution will evolve, if an interdisciplinary team is used to assist in formulating the problem, defining the criteria and parameters of a solution, considering the alternatives, and defining the best potential solution. These should all be done before fixing a mindset. Even a computerized IMS does not make decisions, but rather provides an organized, effective method of processing the large amount of data and information that will be critical in making decisions.

12.6   Dealing with Shrinking Natural Aggregate Sources

For nearly 100 years, the United States has been constructing major civil-engineering works. Most of these civil works use natural aggregates. Aggregates make up 90 to 95 percent of almost any civil-engineering infrastructure project, either alone as an aggregates base or as the major component of asphalt concrete, Portland cement concrete, or vitrified clay. They also form part of any stabilized materials option. In many places in the world the natural aggregates are almost depleted, and there is a strong need to replace natural high-quality aggregate sources with either manmade materials, waste materials, recycled materials, or lower-quality aggregates that heretofore have been rejected under existing specifications.

Historically, civil engineers have incorporated high quality in specifications. Such specifications were presumed to have little or no potential for failure in an absolute sense, but few if any comparisons were made. It has now become necessary to modify standard specifications to accept aggregates that do not meet the extremely high standards used in the past. Such materials can serve well and may cost less in the long run. A methodology for dealing with this concept has been developed by Saeed [Saeed 96].

12.6.1   An IMS subsystem for evaluating materials including waste and reclaimed materials

Not all materials are equal in quality, and quality itself has a wide variety of criteria, ranging from stiffness and initial strength to durability and long-term performance. Energy consumption in the production of new materials and shrinking supplies are all related to the problem. To deal with new materials, as well as waste and reclaimed materials (WRMs), an IMS subsystem has been developed by Saeed [Saeed 96].

Significant pressure is building to use a wider range of materials in infrastructure construction, maintenance, and rehabilitation. This is due to: (1) shrinking funds; (2) limited supplies of high-quality materials near the point of utilization; (3) difficulty in opening new quarries due to zoning laws; (4) increased haul distances and transportation cost; (5) societal concerns and environmental pollution; (6) increased cost of land utilization; and (7) an urgent need to reduce the amount of energy consumed in the production of natural aggregate.

No known method used in design prior to 1996 provided a method for considering all of the diverse factors associated with selecting materials. The stand-alone materials evaluation system [Saeed 95], called the WRM Evaluation System, was developed in a research project for the Texas DOT. It can, however, be applied to all types of infrastructure. The method considers four categories of attributes: (1) technical; (2) economic; (3) societal; and (4) environmental. These come into play after an initial screening process based on local information. The screening process ensures that the proposed material possesses certain minimal fundamental physical, chemical, mechanical, and thermal properties before it is run through the selection subsystem for possible use. The method considers trade-offs between the various strength and performance categories against economic, environmental, and societal issues. While it is difficult to evaluate the societal and environmental issues economically, Saeed provides a method for considering expert opinion and public opinion in this regard. The outline of the method is presented in Figure 12.3.

Enormous amounts of new waste materials are generated in the United States daily. For some agencies and persons, using these waste

Figure 12.3 Outline of method to evaluate factors in selecting materials. [Saeed 95].

materials has become a responsibility, one that can assist in preserving the world's natural resources. Using waste materials can also reduce the world's dependence on natural aggregate [Saeed 95, TGLO 93].

In the technical literature pertaining to use of waste materials, the term *recycling* is often used. Recycling, in essence, is taking an item, after it has been through its useful life, and remaking it into a new, useful product instead of discarding it. *Reusing,* on the other hand, refers to using an item over and over again in its current form without converting it into something new. *Reclaiming,* as defined by Wood et al., is the process during which a waste material is removed from its original location for reuse or recycling at some other location [Wood 89].

Construction, maintenance, and rehabilitation (M&R) of transportation infrastructure is costly and uses large quantities of material. Utilization of WRMs in these activities is expensive, but at times has advantages over the use of conventional materials. Prominent among the major benefits are the conservation of energy and natural resources, as well as preservation of the environment and existing highway geometrics [NCHRP 90].

The typical approach to characterizing WRMs has been to evaluate them technically in the laboratory, and compare the findings with standard specifications for standard natural materials. This is not an effective method because these materials generally do not equal natural aggregates in quality. They do, however, often have a high societal, environmental, and economic value.

There are many reasons for an agency to consider the use of WRMs in transportation infrastructure construction and M&R activities [Han 95]. These include local shortage of natural aggregates, high cost of waste disposal, commitment to the environment, availability of natural and waste materials, political pressure, environmental safety, and others. Cost-effectiveness, performance, availability, and the prevailing political climate are the four issues fundamental to determining if a material is appropriate to use in construction.

An IMS subsystem is needed to objectively evaluate meaningful recommendations. Conclusions can be categorized as technical, economic, societal, and environmental in nature in the subsystem. A simple methodology is needed to evaluate WRMs for potential use in infrastructure construction and M&R activities.

12.6.2   Evaluation subsystem

Any WRM evaluation method must consider the technical, economic, societal, and environmental aspects of their use, after an initial screening process. A technical evaluation unit must evaluate WRMs based on the physical, chemical, mechanical, and thermal properties required of a material for transportation project use. Technical properties determined to be of importance will be investigated in laboratory studies to technically evaluate WRMs.

The economic evaluation unit must identify waste materials that are most viable in an economic sense. This unit must consider all the costs related to the production of construction aggregate from WRMs, and compare these to costs incurred in the production of natural aggregate and the disposal of waste material.

Societal and ecological implications often cannot be measured directly in dollar terms. However, the social evaluation unit should consider the interest or desire expressed by society in general, as well as the government and the private sector.

The environmental impact of using WRMs includes some factors that are related, directly or indirectly, to technical, economic, and social evaluation. The environmental evaluation unit must measure the actual impacts as opposed to how people feel about them.

Objective data is required for any WRM evaluation method in all four evaluation units. This objective data should be supplemented with subjective data in cases where objective information does not exist or is difficult to quantify, such as environmental impact and social values.

12.7 Summary

Infrastructure managers continue to be faced with complex issues in providing the best infrastructure benefits. As outlined in this chapter, dealing with new concepts and materials are on the list. Several examples have been given, although comprehensive details are not included. It is hoped that the examples given here will prompt readers, both managers and operational-level people involved in optimizing infrastructure resources, to use an IMS as a tool to assist in their work. Examples given later in this book also can serve to assist readers.

References

[AASHTO 93] *AASHTO Guide for the Design of Pavement Structures,* American Association of State Highway and Transportation Officials, Washington, D.C., 1993.
[ACI 72] "A Design Procedure for Continuously Reinforced Concrete Pavements for Highways," *ACI Journal,* American Concrete Institute, Detroit, Mich., 1972.
[ACI 84] *Design Handbook: In Accordance with the Strength Design Method of ACI 318-83,* American Concrete Institute, Detroit, Mich., 1984.
[ACI 92] *Concrete Slabs on Grade: Design, Specifications, Construction, and Problem Solving,* American Concrete Institute. Detroit, Mich., 1992.
[AI 87] "Thickness Design-Asphalt Pavements for Air Carrier Airports," *Manual Series No. 11,* Asphalt Institute, 1987.
[AI 91] "Thickness Design-Asphalt Pavements for Highways and Streets," *Manual Series No. 1,* Asphalt Institute, College Park, Md., 1991.

[AISC 94]   *Load and Resistance Factor Design: Manual of Steel Construction,* American Institute of Steel Construction, Chicago, Ill., 1994.

[Bradford 93]   S. A. Bradford, *Corrosion Control,* Van Nostrand Reinhold, New York, 1993.

[Bridge 95]   "Keynote," *Bridge Design and Engineering,* Oct. 1995, pp. 19–20.

[Dawson 95]   D. K. Dawson, "Cast Polymers Capture Sophisticated Markets," *Composites Technology,* Vol. 1, No. 3, Sept./Oct. 1995, pp. 17–20.

[Ferguson 88]   P. M. Ferguson, J. E. Breen, and J. O. Jirsa, *Reinforced Concrete Fundamentals,* John Wiley and Sons, New York, 1988.

[HRB 62]   "The AASHO Road Test Pavement Studies," *HRB Special Report 61D,* Highway Research Board, National Academy of Science, Washington, D.C., 1962.

[Han 95]   C. Han, and A. M. Johnson, "Waste Products in Highway Constructions," *Proceedings, Sixth International Conference on Low-Volume Roads,* Transportation Research Board, Washington, D.C., 1995.

[Iyer 95]   S. L. Iyer, "Design and Construction of FRP Cable Prestressed Navy Pier at Port Hueneme, California (CPAR Program)," *Proceedings, Fiber Reinforced Plastics Workshop,* sponsored by the Federal Highway Administration and TRB Committee A2C07, Jan. 1995.

[Jacobs 94]   J. A. Jacobs and T. F. Kilduff, *Engineering Materials Technology,* Instructor's Manual, 2nd ed., Prentice-Hall, Englewood Cliffs, N.J., 1994.

[Jones 93]   M. A. Jones, "Cost Differences between Expediting and Standard Urban Freeway Construction," Master's thesis, the University of Texas at Austin, Dec. 1993.

[McCullough 60]   B. F. McCullough and W. B. Ledbetter, "LTS Design of Continuously Reinforced Concrete Pavements," *Journal Highway Division,* vol 86. No. HW4, Dec. 1960.

[McCullough 68]   B. F. McCullough and H. J. Treybig, "A State Deflection Study of Continuously Reinforced Concrete Pavement in Texas," *Highway Research Record 239,* Highway Research Board, National Research Council, Washington, D.C., 1968.

[NCHRP 90]   J. A. Epps, *Synthesis of Highway Practice Report 160: Cold Recycled Bituminous Concrete Using Bituminous Material,* National Cooperative Highway Research Program, Transportation Research Board, National Research Council, Washington, D.C., 1990.

[PCA 84]   *Thickness Design for Concrete Highway and Street Pavements,* Portland Cement Association, Skokie, Ill., 1984.

[Plaxico 95]   C. A. Plaxico, R. M. Hackett, and W. Uddin, "A Micromechanical Analysis of the Viscoelastic Response of Resin Matrix Composites," *Proceedings, International Conference on Fibre Reinforced Structural Plastics in Civil Engineering,* Madras, India, Dec. 1995, pp. 99–108.

[Plaxico 96]   C. A. Plaxico, W. Uddin, and R. M. Hackett, "A Micromechanical Model for Asphalt Materials," *Proceedings, Fourth Materials Congress, American Society for Civil Engineers,* Washington, D.C., Nov. 1996, Vol. 1, pp. 761–770.

[Saeed 95]   A. Saeed, W. R. Hudson, and P. Anaejionu, "Location and Availability of Waste and Recycled Materials in Texas and Evaluation of their Utilization Potential in Roadbase," *Research Report 1348-1,* Center for Transportation Research, the University of Texas at Austin, Austin, Tex., Oct. 1995.

[Saeed 96]   A. Saeed and W. R. Hudson, "Evaluation and Use of Waste and Reclaimed Materials in Roadbase Construction," *Research Report 1348-2F,* Center for Transportation Research, the University of Texas at Austin, Austin, Tex., Aug. 1996.

[Seible 95]   F. Seible and D. Innamorato, "Earthquake Retrofit of Bridge Columns with Continuous Carbon Fiber Jackets," *Report No. ACTT-95/08,* Vol. II: *Design Guidelines, Report to Caltrans,* Advanced Composites Technology Transfer Consortium, University of California, San Diego, Calif., Aug. 1995.

[Sohail 95]   F. Sohail, "Implementation of the Urban Roadway Management System," Master's thesis, the University of Texas at Austin, Aug. 1995.

[Solminihac 92]   H. de Solminihac, "System Analysis for Expediting Urban Highway Construction," Ph.D. thesis, the University of Texas at Austin, May 1992.

[Sultan 95] M. Sultan, "Caltrans Program for Performance Testing of Fiber Reinforced Plastics for Seismic Retrofit and Rehabilitation of Structures," *Proceedings, Fiber Reinforced Plastics Workshop,* sponsored by Federal Highway Administration and TRB Committee A2C07, Jan. 1995.

[Technology 95] "Naval Test Pier Gets Composite Catwalk," *Composites Technology,* Vol. 1, No. 3, Sept./Oct. 1995, pp. 5–6.

[TGLO 93] G. Mauro, *Texas Recycles: Marketing Our Neglected Resources,* Texas General Land Office, Austin, Tex., 1993.

[USACE 58] "Engineering and Design-Flexible Pavements," *EM-1110-45-302,* Waterways Experimental Station, U.S. Army Corps of Engineers, 1958.

[USACE 61] "Revised Method of Thickness Design of Flexible Highway Pavements at Military Installations," *Technical Report No. 3-582,* Waterways Experimental Station, U.S. Army Corps of Engineers, Aug. 1961.

[Wood 89] L. E. Wood, T. D. White, and T. B. Nelson, "Current Practice of Cold in-Place Recycling of Asphalt Pavements," *National Seminar on Asphalt Pavement Recycling,* Transportation Research Board, National Research Council, Washington, D.C., 1989.

Economics, Life-cycle Analysis, and M,R&R Programming

Maintenance, Rehabilitation, and Reconstruction (M,R&R) Policies and Treatment Alternatives

13.1 Introduction

The strategic aspect of M,R&R (maintenance, rehabilitation/renovation, and replacement/reconstruction) was presented in Chapter 11. In essence, the term *strategic* refers to a planned approach. This includes a definition of what is meant by maintenance, rehabilitation, renovation, and reconstruction, and a description of the major elements of such a strategic approach or direction. Included are the concepts of maintainability, explicit consideration of future maintenance and rehabilitation at the design stage, reliability-centered maintenance, maintenance management systems, and ongoing operations.

By comparison, the term *policy* involves the set of rules and guidelines required to realize the strategy set out by an agency in regard to the way work will actually be done. This includes the following elements:

- Creating an operational form of the maintenance, rehabilitation, or overall infrastructure asset management system

- Determining whether the work should be contracted out or done by in-house forces

- Determining or specifying an appropriate policy, such as preservation (i.e., preventive maintenance and timely rehabilitation) or a reactive type of policy (i.e., dealing only with corrective actions or problems as they arise)

- Identifying the M,R&R treatment alternatives to be considered, and in turn identifying those that are feasible for a given set of conditions
- Defining procedures for evaluating the effectiveness of M,R&R treatments
- Defining the collection and reporting of M,R&R data
- Explicitly recognizing the importance of environmental impacts and policies

13.2   Maintenance Management Systems

An overall framework for infrastructure management is outlined in Chapter 1 (see Figure 1.9). It can apply to any infrastructure asset. More particularly, it is useful to review the operational form of management systems within the overall framework. One approach, which has existed for several decades, is the maintenance management system (see also Sec. 11.7). In effect, most maintenance management systems are a form of the production control process, especially in the roads and highway sector of infrastructure management. Figure 13.1 presents the basic operational form or functional extent of most such maintenance management systems.

The functional extent of maintenance management systems represented in Figure 13.1 has existed relatively unchanged for 25 years or more. A study by the U.S. National Cooperative Highway Research Program [NCHRP 94] has noted, however, the following:

> The current maintenance management systems were right for their time, helping in the planning, budgeting, monitoring, and evaluation of maintenance work, and fostering standard methods and productivity guidelines. Many changes have occurred since these systems were first developed, however; in this light, several improvements are needed in current systems.

The NCHRP study goes on to point out that most maintenance management systems use a one- to two-year outlook rather than multi-year planning. They employ a centralized approach to planning, scheduling, and control that can be burdensome for field personnel, and they are not flexible enough to adjust work plans and schedules to reflect changing conditions. They do not incorporate distinct measures of levels of service or quality standards, and they do not have the capability to adjust resource requirements based on the degree of out-sourcing to be used. Finally, NCHRP reports that current MMS are not able to generate timely reports to various management levels. All of these deficiencies should be addressed in future maintenance

Figure 13.1   Basic operational form of most maintenance-management systems.

management systems, preferably as a synchronous part of an IMS. In fact, the NCHRP report specifically recommends that the next generation should build upon the experience gained from the integration of engineering, economic, and management principles in pavement management systems.

Another source of information relevant to the operational form that maintenance and/or other infrastructure management systems might consider are in the American Public Works Association book, *Public Works Management Practices*. It recommends criteria and procedures for public-works management services [APWA 91].

13.3   Privatizing/Contracting-out M,R&R

It has been widespread practice in the U.S., Canada, and other countries to contract out rehabilitation and reconstruction work in almost all areas of infrastructure. By comparison, however, it has been widespread practice to use in-house forces for routine maintenance of roads, water, sewer, buildings, parks, and utilities. The general rationale for the latter approach has been historical (i.e., things have always been done that way). The perception was that permanent teams could be more responsive to short-term, immediate needs. Although some contended that it was more cost-effective, in many cases it has been a convenient mechanism for awarding political jobs and thus gaining votes at the local level.

In recent years there has been a growing shift to the privatization of maintenance, particularly in the highway sector but also involving other areas of infrastructure, such as building maintenance. A noteworthy and very ambitious initiative comes from Australia, where a 10-year total road maintenance contract was awarded in late 1995 involving 2,000 lane-kms in the northeastern region of Sydney [TEACC 96]. It covers pavements, nearly 200 bridges, over 600 traffic signals, guard rails, right-of-way, curbs and gutters, signs, and drainage systems (i.e., everything within the right-of-way). Maintenance, in this case, has a very broad definition, ranging from routine, reactive type of work, such as pothole repairs, to preventive and corrective work to reconstruction. The contract is performance-based in that certain long-term performance indicators and asset preservation targets have to be met. It represents a true partnership between the contractor and the Road Transport Authority (RTA). The contract price is $170 million (Australian dollars), with periodic lump-sum payments to the contractor.

13.4   Identifying M,R&R Alternative Treatment Policies

Pursuant to discussions in Chapter 11, M,R&R policies need three ingredients: (1) M,R&R intervention-level criteria; (2) M,R&R alternative strategies with unit cost and production data; and (3) life estimates of each strategy or an appropriate performance model. The selection of an appropriate policy depends on a number of technical and economic factors. Broadly speaking, the following categories of policies are included.

13.4.1   Do-nothing policy

A do-nothing policy implies that no significant maintenance is done to the facility, and the condition deterioration continues until the time at which the facility is abandoned, reconstructed, or replaced. This may be a viable policy if a new facility is planned to replace the existing one (to satisfy a growing need or demand), or if budgeted rehabilitation/restoration methods cannot extend the life to the desired number of years. A do-nothing policy can be used as a base case to calculate the benefits of other alternatives. It also works well if replacement is funded within three years.

13.4.2 Routine maintenance policy

Routine maintenance policy generally involves periodic actions to preserve the condition and service levels. It is usually programmed as an annual lump-sum expenditure. Example actions are periodic mowing; general cleanup of common areas, parks, and roads; and regular maintenance and cleanup of electrical and mechanical equipment, such as elevators, pumps, and air-conditioning units. This is sometimes termed *preventive maintenance*.

13.4.3 Critical maintenance policy

Critical maintenance involves a policy of immediate action(s) to avoid imminent failures and emergency situations resulting from accidents and natural disasters. It also includes actions responding to the user or the public's complaints. Emergency repairs are performed in the case of structural damage and failures caused by accidents (for example, derailment of railroad cars, spillage of hazardous materials, gas pipeline blowups, and failure of underground water pipes) and natural disasters (for example, collapse of bridges and buildings during earthquakes, land slides, floods, tornadoes, and hurricanes).

13.4.4 Scheduled M,R&R policy

Scheduled M,R&R includes both timed maintenance and replacement/reconstruction. This type of policy is particularly applicable when past experience exists (for example, painting of steel bridges, pavement marking, exterior painting of a building, replacement of gutters of residential buildings, and scheduled maintenance of pumps and auxiliary equipment of a water-supply system). This type of policy would also include: (1) hard-time replacement (for example, the shingle roof of a residential building may be replaced after 20 years because of the known average life of this roofing material); and (2) cost-effective capital replacement/reconstruction strategies (for example, asphalt resurfacing of a bridge deck every 15 years). This differs from preventive maintenance since it is longer-term, usually five or more years.

13.4.5 Condition-responsive M,R&R policy

Condition-responsive M,R&R policy is based on preselected threshold values(s) of appropriate performance indicators that will trigger major rehabilitation/renovation and replacement/reconstruction actions. It is important to carefully identify these performance indicators. Examples would be frequency of breaks, amount of leakage in water pipelines, or level of scour near a bridge pier. It is very helpful to have good predictive performance models for timely budgeting.

13.4.6   Database of M,R&R actions

A database of all M,R&R actions taken under the foregoing policies, or for any reason, is essential for updating the maintenance programs and evaluating their effectiveness. At a minimum, the database should include M,R&R activity codes, timing of the work, units of measure, unit costs, expected life, production rates, and condition improvement after implementation. Additionally, it is desirable to record a condition evaluation before and after each action. In the form of a historical record, the database also is valuable for performance-modeling purposes (see Chapter 8).

13.4.7   Economic analysis of M,R&R policies

Economic analysis of competing M,R&R strategies on a life-cycle basis is discussed in detail in Chapter 14. At this point, the importance of comparing the overall economics of the foregoing policies is noted. Such comparisons can be based on the more detailed analysis of competing strategies and may reveal, for example, whether a scheduled or condition-responsive M,R&R policy is more cost-effective for a given situation.

13.5   Example M,R&R Treatment Alternatives for Roads

Haas, Hudson, and Zaniewski [Haas 94] present detailed examples of M,R&R strategies and analysis methods for pavements. Maintenance treatments depend on the pavement surface type and its use, predominant distress, and deterioration mechanisms [Haas 94, Roberts 91]. Long-term performance of various maintenance strategies is being evaluated in the Strategic Highway Research Program (SHRP), and is periodically reported in the literature. For flexible pavements, surface treatments, thin asphalt overlays, and crack sealing are generally used as global preventive maintenance treatments, while patching is used as a localized, corrective maintenance treatment. Major rehabilitation alternatives include hot-mix asphalt overlays, concrete overlays, surface milling and recycling, and reconstruction. For rigid pavements, cleaning and rehabilitation of joints, slab replacements, patching, and thin asphalt overlays are generally used as preventive maintenance treatments. Rehabilitation alternatives include hot-mix asphalt overlays, concrete overlays, and reconstruction.

Noise generation on heavily trafficked roads can force the use of special treatments, such as sound-barrier walls erected on both sides of the right-of-way. Innovations in thermoplastics are providing new materials that could prove to be maintenance-free, show relatively higher sound-absorption characteristics (as compared to conventional concrete materials), and reduce the wall height. An example is the application of polycarbonate material to reduce the sound-wall heights on the I-5 freeway in Los Angeles [Gharabegian 96].

13.6 Example M,R&R Alternatives for Water Mains

An analysis of the deterioration and maintenance problems of aging water-supply systems has identified three key areas of M,R&R planning [O'Day 84]: (1) leakage; (2) main breaks; and (3) risk analysis. Utilities must recognize the risk of a main failing, and the risk of damage and loss from the failure. Such damage can result in floods, disruption of traffic, fire, and substantial losses to affected businesses. In general, a break in a 122-cm (48-in) main will cause more loss and damage than several breaks in a 15-cm (6-in) main. There are three basic types of commonly-used M,R&R alternatives or actions: general maintenance (main cleaning, equipment repair, valve service, treatment-facility cleaning), repair (main lining, main repair), and replacement (main relaying). Table 13.1 lists the major deterioration problems and maintenance actions reported for water systems.

TABLE 13.1   Water-Distribution System Problems and Maintenance Actions [O'Day 84]

Problems/deterioration	M,R&R action
Water quality affected by tuberculation and internal pipe corrosion	Main cleaning and lining, cathodic protection, facility maintenance (preventive and corrective maintenance)
Low pressure and high head loss due to main tuberculation	Main cleaning and lining (preventive maintenance)
Valves and hydrants in poor condition or inoperable due to neglect and/or unavailable replacement parts	Equipment maintenance (routine and scheduled maintenance)
Distribution system leakage due to joint leaks, main breaks, and service leaks	Leakage survey and repair, service leak inspection and removal of abandoned service (condition-responsive/preventive maintenance)
Main breaks due to main deterioration from internal and/or external corrosion	Repair of replacement of main breaks, main relaying

Water-main repair or replacement decisions become critical when mains deteriorate structurally and break frequently. In-service evaluation of water mains is discussed in Chapter 7. The following evaluation criteria can be used to make replacement decisions [Stacha 68]:

- Projected future costs of continued repair
- Initial cost of main replacement
- Public risk from future breaks
- Frequency and severity of inconveniences to the public from future breaks
- Hydraulic adequacy of the main
- Future plans for streets and/or utility construction that could affect break patterns and alter costs through joint construction
- Severity of disruptions caused by main-replacement construction activity

An innovative preventive maintenance treatment is the use of in situ lining to improve leakage control and repair existing mains, such as the Insituform process [News 94]. This lining employs thermoplastic technology for rebuilding deteriorated pressure pipes, without relying on the structural properties of the pipe. Economic analyses of leak control and repair versus replacement of main breaks are discussed in Chapter 14.

13.7 Example M,R&R Alternatives for Sewer Mains

Urban water-supply and waste-water systems are regulated in the U.S. by the Clean Water Act. Sewers collect waste water from residential communities and businesses. Control of pollution is achieved by the proper treatment of waste water. Maintenance of filter systems in drainage basins, trenches, and wells is necessary on a periodic basis. Catch basins should be inspected and cleaned after major storms. Corrugated steel pipes for sewer lines and culverts have been used for 100 years [AISI 91]. The life of these pipes can be extended by 25 percent with external bituminous coating, and 15 to 35 years of service life can be added if invert bituminous or concrete paving is applied to the bottom 25 percent of the pipe circumference. The most frequently used equipment for infiltration maintenance includes vacuum pumps to clean sumps and pipes, compressed air jets to clean wells, and flushing. Methods of rehabilitating deteriorated steel pipes include installation of a concrete invert for large diameters, relining with smaller-diameter pipe, shotcrete relining, and cement mortar relining [AISI 90].

TABLE 13.2   WSSC Cost Comparison of Sewer Reconstruction Techniques
[Urban 84c].

Type of job	Excavation/ replacement cost per foot	Slip lining estimated cost per foot	Insituform actual cost per foot
Sewer line, 25.4-cm (10-in) diameter and 88.5-m (295-ft) long; 100% of the work involved paving replacement, and five service laterals were reconnected.	$222	$150	$52
Sewer line, 15-cm (6-in) diameter and 96-m (320-ft) long; 100% of the work involved paving replacement, and nine service laterals were reconnected.	$80	$150	$59
Sewer line, 25.4-cm (10-in) diameter and 684-m (2280-ft) long; 27% of the work involved paving replacement, and no service laterals were involved.	$69	$60	$52
Sewer line, 25.4-cm (10-in) diameter and 428-m (1425-ft) long; 80% of the work involved paving replacement, and six service laterals were reconnected.	$84	$80	$52

Like water mains, sewer mains can also be lined with new thermo-plastic materials to provide an acid-alkali-resistant and structurally durable rehabilitation/repair alternative [News 94]. The Washington Suburban Sanitary Commission (WSSC), responsible for providing water and sewer services to Montgomery and Prince George counties in Maryland, compared the costs of replacement, slip lining, and Insituform for 15-cm (6-in) and 25.4-cm (10-in) sewers in a number of projects. The result was significant savings using the newer "trench-less" technology of Insituform, which involved no excavation [Thomasson 82, Urban 84c]. These comparisons are summarized in Table 13.2.

13.8 Example M,R&R Alternatives for Buildings

Buildings generally have more components than other types of infra-structure. The effective life of a building will end or be greatly reduced if one or more of these components (structure, foundation, mechanical, roof) fails. The finishes, fixtures, and furnishing may be critical for the owner or user because of the visual, decorative aspects; however, the architectural and engineering condition assessment is usually focused on functional and structural integrity. Harris defines six functional subsystems of a building that account for the service provided, the individual failure of any of which can lead to overall functional failure [Harris 96]. These six subsystems are listed as follows:

1. *Structure*—to minimize deformation and prevent cracking. This includes foundations, compacted ground, earth-retaining structures, bearing walls, columns, beams, girders, trusses, and braces.

2. *Vertical closure*—to provide weather barrier. This includes nonstructural components of exterior walls, siding, windows, doors, and sills.

3. *Horizontal closure*—to collect and transport precipitation. This includes roofing materials, gutters, down-spouts, drains, retention ponds, sump pits and pumps, and sewers.

4. *Climate stabilization*—to maintain desirable levels of heat, humidity, and oxygen. This includes heating, ventilation, air conditioning, ducts, controls, fireplaces, insulation, fans, pumps, motors and other necessary equipment.

5. *Hydraulic*—to distribute water and collect and dispose of sewer. This includes water mains, water-distribution system, plumbing, grease traps, water heaters, wells, pumps, storage tanks, drains, valves, vents, sanitary sewers, septic tanks, and hydrant and fire protection.

6. *Energy distribution*—to distribute power and collect by-products. This includes fuels and energy of all types and return of wastes, electrical installation, wiring, meters, gas lines, oil tanks and pumps, and pressure regulators and controls.

Obviously, there are many potential M,R&R alternatives to address failure or deterioration of the above functional subsystems. The reader may consult various industry and public-agency documents for details, particularly the publications of the Building Research Board, National Academy of Science, Washington, D.C.

13.9   Evaluating the Effectiveness of M,R&R Alternatives

The evaluation of effectiveness of M,R&R treatment alternatives can be carried out in one or more of the following ways:

- Economic effectiveness
- Technical effectiveness
- Visual, or aesthetic, effectiveness
- Public or owner satisfaction/acceptance
- Environmental impact

Economic effectiveness is usually evaluated on a life-cycle basis in terms of uniform annual costs, as described in Chapter 14. Technical effectiveness is usually evaluated in terms of performance and service to the users (see Chapter 8 on performance modeling), and is thus an integral part of any life-cycle-based economic-effectiveness evaluation.

It is sometimes important to evaluate the visual, or aesthetic, effectiveness of roads, bridges, retaining walls, parks, buildings, and various other types of infrastructure that are observed by users or owners. The most simple approach might consist of a quality and appearance scale, such as: (1) excellent; (2) good; (3) fair; (4) poor; and (5) unacceptable. More sophisticated approaches could also be used, but the determining factor would be the degree of importance that visual/aesthetic effectiveness plays in the overall effectiveness evaluation. This value could range from essentially zero for underground services to high for a retaining wall. Clearly, park and public-service facilities, such as museums, find this important, as do schools.

A similar situation exists for public or owner satisfaction. This is quite a general category, and may include economic, technical, and visual effectiveness, depending on the owner. Very few IMSs currently evaluate such overall satisfaction.

Environmental-impact effectiveness is usually addressed through an environmental-impact assessment prepared according to applicable federal, state, or local regulations. If the standards incorporated in the regulations are met, then the infrastructure is generally environmentally acceptable. In other words, it is acceptable or not acceptable. Of course, if acceptable, there can be a range of effectiveness within that category that can be considered.

13.10   Collection and Reporting of M,R&R Data

It was noted in Sec. 13.1 that a comprehensive policy approach includes not only the necessity of collecting and reporting M,R&R data, but also how this is carried out. Chapters 4, 5, and 6 have addressed the details of the data needs, data analysis, database management, and procedures for collecting and processing data. It is not intended herein to repeat these details. Rather, it is intended to emphasize the importance of ensuring that an agency policy would include the capability of determining the following fundamental information (see also Sec. 13.4):

- The M,R&R actions or treatments that were carried out, in what locations, when, and to what degree/quality/extent

- The costs, problems encountered, and expected life of the infra-structure

In addition to the above, a variety of data may be required to assess effectiveness and develop performance models, as discussed in Chapters 4, 5, 6, and 8.

13.11 Recognizing Environmental Impact/and Policies

Most M,R&R actions must conform to environmental regulations or policies that require that the impacts of each action be considered. Exceptions would be minor repairs or renovations.

Usually, the impact assessments include possible disturbances to the natural environment, creation of bad by-products and emissions, disturbances to manmade works, creation of noise, and effects on public health. As previously indicated, it is essential to check with the relevant environmental office (federal, state, local) to determine which environmental requirements have to be involved. Generally, most governmental agencies have a regular channel of environmental coordination.

13.12 Summary

All possible M,R& R treatment possibilities for all types of infrastructure cannot be covered in this chapter. Instead, this chapter illustrates the need to consider a range of possibilities, including new technology. One should not get stuck in a rut of outdated methods; approaches should be evaluated and assessed as outlined for the specific type of infrastructure involved. In this way, a more complete and effective IMS can be developed to provide the maximum benefits for available funds.

References

[AISI 90]   *Modern Sewer Design,* American Iron and Steel Institute (AISI), Washington, D.C., 1990.

[APWA 91]   *Public Works Management Practices,* American Public Works Association (APWA), Chicago, Ill., 1991.

[Gharabegian 96]   A. Gharabegian, "Soundwall Extension Quiets Noise along Los Angeles Freeway," *Roads and Bridges,* April 1996, pp. 71–72.

[Haas 94]   R. Haas, W. R. Hudson, and J. P. Zaniewski, *Modern Pavement Management,* Krieger Publishing Company, Malabar, Fla., 1994.

[Harris 96]   S. Y. Harris, "A Systems Approach to Building Assessment," *Standards for Preservation and Rehabilitation, ASTM STP 1258,* American Society for Testing and Materials, Philadelphia, Pa., 1996, pp. 137–148.

[News 94]   "Buyer's Guide," *Civil Engineering News,* Vol. 6, No. 2, March 1994, p. 43.

[NCHRP 94]   National Cooperative Highway Research Program, "Role of Highway Maintenance in Integrated Management Systems," *NCHRP Report 363*, Transportation Research Board, National Research Council, Washington, D.C., 1994.

[O'Day 84]   D. K. O'Day, "Aging Water Supply Systems: Repair or Replace," *Infrastructure—Maintenance and Repair of Public Works, Annual of the New York Academy of Science*, Vol. 431, Dec. 1984, pp. 241–258.

[Roberts 91]   F. L. Roberts, P. S. Kandhal, E. R. Brown, D.-Y. Lee, and T. W. Kennedy, *Hot Mix Asphalt Materials, Mixture Design, and Construction*, NAPA Education Foundation, Lanham, Md., 1991.

[Stacha 68]   J. H. Stacha, "Criteria for Pipeline Replacement," *Journal of American Water Works Association*, May 1968.

[TEACC 96]   "The New Force in Road Maintenance," *The Earth Mover and Civil Contractor*, Aug. 1996.

[Thomasson 82]   R. O. Thomasson, "In-Place Sewer Reconstruction Proves Cost-Effective," *American City and County*, Feb. 1982.

[Urban 84c]   H. P. Hatry and B. G. Steinthal, *Guides to Managing Urban Capital Series*, Vol. 4: *Guide to Selecting Maintenance Strategies for Capital Facilities*, The Urban Institute, Washington, D.C., 1984.

14

Life-cycle Cost and Benefit Analysis

14.1 Introduction

Applying the principles of economic analysis to infrastructure projects occurs at two basic levels. First, the overall economic viability and timing of a project must be determined. This is most often carried out in comparison to and/or competition with other projects. Second, there is the requirement to achieve maximum economy for a project once it has been selected, programmed, and budgeted. Such within-project economic analysis is achieved by calculating the costs and benefits of the various alternatives that satisfy overall project requirements. In terms of economic analysis, the major difference between these two levels of infrastructure management is the amount of detail or information needed. Otherwise, the basic principles are the same.

This chapter describes both the economic principles and their incorporation into methods of economic analysis. Such analysis is a vital component of any infrastructure management system [Army 86, DOI 83, TRB 85]. An example of life-cycle benefit and cost analysis, involving a case study, is provided to illustrate the principles and the methodology. More complete analysis details can be obtained in any good engineering economics text.

14.2 Basic Principles

Considerable information exists on the principles of engineering economy and the methods of economic analysis. The principles applicable to infrastructure management include the following:

- The level at which the analysis is to be made must be clearly identified, i.e., expenditure programming for a number of projects or network, or within project optimization.

- The economic analysis provides support for a management decision, but does not alone represent a decision.

- Criteria, rules, or guides for such decisions must be separately formulated before the economic analysis, even though such criteria may be straightforward and simple (i.e., minimization of life-cycle costs).

- The economic analysis itself has no relationship to the financing of a project. Such financing considerations can either limit the number of feasible projects (on a network-level basis) or limit the amount available for a particular project. They do not affect the methodology or principles controlling the economic analysis per se.

- An economic analysis should consider as many feasible alternatives as possible, within the constraints of time and other resources.

- All alternatives should be compared over the same life-cycle or time period. The time period should be chosen so that it is possible to forecast the factors involved in the analysis with a reasonable degree of reliability. Uncertainties can be considered in making the decision to choose the best alternative.

- The economic analysis of infrastructure projects should include agency costs and user costs and benefits, if possible.

14.3   Cost and Benefit Factors

14.3.1   Agency costs

Agency costs are those costs directly represented by the budget or out-of-pocket costs paid by the owner. External or exogenous costs, such as those associated with user delays during infrastructure work, air pollution, and the like (usually termed *user costs*), may be combined with agency costs in a total economic analysis, and they may or may not affect agency decision making, but they do not appear in an agency's budget.

The major initial and recurring costs over the life-cycle that a public agency may consider in the economic analysis of infrastructure project alternatives include the following:

- Initial capital costs of construction

- Future costs of maintenance, rehabilitation, renovation, and reconstruction (M,R&R)

- Salvage return or residual value at the end of the period (which may be a "negative cost")
- Engineering and administration
- Costs of borrowing (if projects are not financed from current revenue)

14.3.2 Nonagency costs

Nonagency costs can involve the user of the infrastructure or facility, or they can be incurred by nonusers. They include the following:

- User Costs
  1. Occupancy time in or on the facility
  2. Operating costs (vehicles, etc.)
  3. Accidents
  4. Time delays due to M,R&R activities
- Nonuser Costs
  1. Environmental pollution (emissions, noise, visual, etc.)
  2. Neighborhood disruptions

Some of the above costs are difficult to quantify. In the transport field, however, considerable progress has been made on quantifying user time and time-delay costs plus operating costs [TRB 85, Uddin 93]. For nonuser costs, most work in the transport field has focused on emissions and noise.

14.3.3 Benefits

The benefits of an infrastructure project can accrue from direct or indirect cost reductions, and from gains in business, land use and values, aesthetics, and community activities in general. Benefits would, however, accrue primarily from direct reductions in user costs, as listed in the preceding section. It is also possible to consider benefits in terms of additional taxes or total income generated by a project.

To measure or calculate benefits, it is necessary to define those facility or infrastructure characteristics that will affect the previously noted user costs of time, accidents, operating costs, and so forth. These could include level of service, appearance, or aesthetics. However, the first factor is level of serviceability, which can be measured in a number of different ways, depending on the infrastructure component involved, and would have a major influence.

In assigning benefits for the purpose of project evaluation, the question of whether generated or transferred usage of the facility should be included must be considered. Usually, this question arises with respect to the overall project.

This question can relate to the public-policy viewpoint adopted [Winfrey 69]. One viewpoint involves a "sales concept," in which all usage is considered regardless of sources; the other involves a concept of true savings, in which only existing plus normal usage growth over the analysis period is considered. It is usual in transport projects to use the first concept. Thus, including total expected usage in the calculations for costs and benefits for infrastructure improvement projects is not unreasonable.

14.4   Analysis or Life-cycle Period

A general guideline for selecting the length of an analysis period is that it should not extend beyond the period of reliable forecasts. For short-term or high-uncertainty projects, 10 years could be an upper limit. For other long-term projects, such as bridges or structures, 50 years or more may not be unreasonable; however, the present worth of costs or benefits at future times beyond 50 years may not be significant, depending on the discount rate used. Most transport studies use a range of 20 to 30 years. An analysis period of 50 years is used for sewer pipeline materials. The analysis period can also be based on service-life estimates. Examples of service life are provided in Chapter 3 (Tables 3.1 and 3.2) and Chapter 8 (Table 8.1). The particular period chosen is basically a policy decision for the agency concerned, and can vary.

14.5   Discount Rate, Interest Rate, and Inflation

Discount rate, or time value of money, must be used to relate estimated future costs and/or benefits to present worth terms. It provides a convenient and understandable means of comparing alternative uses of funds. However, discount rate may not be the same as interest rate, which is associated with actually borrowing or investing money (functional costs).

Inflation is of concern to everyone. But whether and how it should be taken into account on an economic analysis is a complicated issue. Basically, the difference between an interest rate and an inflation rate represents the real return on an investment; consequently, it is reasonable to use that difference as a discount rate, which is the policy of many agencies. For example, in 1996, interest rates in the U.S. and Canada were at about 5.5 to 6 percent, with inflation at about 1.5 to 2 percent, giving a difference or average discount rate of about 4 percent. This rate is also the approximate long-term growth of the North American economy. Of course an agency may decide, as a matter of policy, to use a higher or lower discount rate. As well, some agencies look at a range of discount rates to determine if this has any effect on the comparative rankings of investment alternatives.

For normal IMS economic analyses, inflation should only be used in setting a real discount rate, but not directly as a separate factor in economic analysis for the following reasons:

- Inflation is difficult to forecast and merely introduces another uncertainty into the evaluation. If it were considered, benefits as well as costs would be increased by inflation so that their relative magnitude would still be the same.

- The purpose of economic analysis is to provide a basis for decision-making. Inserting a factor for inflation is no guarantee that the results will be more reliable, or that a better selection of alternatives will result.

- The uninflated value of the alternatives in real constant dollars is a better tool for economic analysis than inflated values. It also has more intrinsic meaning to current decision-makers.

Some engineers argue that it is necessary to include inflation in an economic analysis, since ignoring it leads to underestimating costs, and proposed budgets may be incorrect. This argument misunderstands the objective of an economic analysis, which is to provide a tool for the selection of a specific option from a set of alternatives. Once the option is selected, a separate budget and financial analysis is required to determine the cash-flow requirements. The budget analysis generally includes inflation, as well as actual sources of funds, such as taxes, federal grants, bonds, and financial analysis.

14.6 Salvage, or Residual Value

Salvage, or residual, value is used by some agencies in economic analysis. It can be significant in the case of buildings, roads, and pavement structures because it involves the real worth of reusable materials, or additional remaining service life at the end of the life-cycle period. With depleting resources, material values will become increasingly important in the future, especially when used in a new application by reworking or reprocessing.

Salvage value of a material or structural component depends on several factors, such as volume and location, degree of contamination, age, durability, and anticipated use at the end of the life-cycle period. It can be represented as a percentage of the original cost of the new material or component itself, not including the incremental cost of placement and processing.

14.7   Methods of Economic Analysis

A number of methods of economic analysis are applicable to the evaluation of alternative infrastructure investment strategies. They can be categorized as follows:

- Equivalent uniform annual cost method, or simply the annual cost method
- Present worth method for
  1. Costs, or
  2. Benefits, or
  3. Benefits minus costs, usually termed "the net present value method"
- Rate-of-return method
- Benefit-cost-ratio method
- Cost-effectiveness method

14.7.1   Equivalent uniform annual cost method

This method combines all initial capital costs and all recurring future expenditures into equal annual payments over the analysis period. In functional form, the method may be expressed as follows:

$$AC_{x1,n} = crf_{i,n}(ICC)_{x1} + (AAMO)_{x1} + (AAUC)_{x1} - sff_{i,n}(SV)_{x1,n} \quad (14.1)$$

where $AC_{x1,n}$ = equivalent uniform annual cost for alternative $x1$, for a service life or analysis period of n years;

$\quad crf_{i,n}$ = capital recovery factor for interest rate i and n years = $i(1 + i)^n/[(1 + i)^n - 1]$;

$\quad (ICC)_{x1}$ = initial capital costs of construction;

$\quad (AAMO)_{x1}$ = average annual maintenance, plus operation costs for alternative $x1$;

$\quad (AAUC)_{x1}$ = average annual user costs for alternative $x1$, if applicable;

$\quad sff_{i,n}$ = sinking fund factor for interest rate i and n years = $i/[(1 + i)^n - 1]$; and

$\quad SV_{x1,n}$ = salvage value, if any, for alternative $x1$ at the end of n years.

The basic appeal of the equivalent uniform annual cost method is its simplicity and ease of understanding for public officials. It also permits the comparison of alternatives that do not have the same overall expected life. However, it cannot be used, except intuitively, to determine whether or not a project is economically justifiable because it does not include benefits in the evaluation. Consequently, comparisons among alternatives must be on the basis of costs alone, with the inherent assumption that they have equal benefits. A bridge to "serve" the public can fit into this category.

The present worth of costs method is directly comparable to the equivalent uniform annual cost method.

14.7.2 Present-worth method

The present-worth method can consider costs only, benefits only, or costs and benefits together. It involves the discounting of all future sums to the present, using an appropriate discount rate. The factor for discounting either costs or benefits is:

$$pwf_{i,n} = 1/(1 + i)^n \tag{14.2}$$

where $pwf_{i,n}$ = present worth factor for a particular i and n;
i = discount rate; and
n = number of years to when the sum will be expended or saved.

The present-worth method for costs can be expressed in terms of the following equation:

$$TPWC_{x1,n} = (ICC)_{x1} + \sum_{t=0}^{n} \left\{ pwf_{i,t} \left[(CC)_{x1,t} + (MO)_{x1,t} + (UC)_{x1,t} \right] \right\} - pwf_{i,n}(SV)_{x1,n} \tag{14.3}$$

where $TPWC_{x1,n}$ = total present worth of costs for alternative $x1$, for analysis period of n years;
$(ICC)_{x1}$ = initial capital costs of construction, etc., for alternative $x1$;
$(CC)_{x1,t}$ = capital costs of construction, etc., for alternative $x1$, in year t, where $t < n$;
$pwf_{i,t}$ = present worth factor for discount rate, i, for t years = $1/(1 + i)^t$;
$(MO)_{x1,t}$ = maintenance plus operation costs for alternative $x1$ in year t;
$(UC)_{x1,t}$ = user costs, if applicable for alternative $x1$ in year t; and

$(SV)_{x1,n}$ = salvage value, if any, for alternative $x1$, at the end of the analysis period, n years.

The present worth of costs is used in the equivalent uniform annual cost method when additional capital expenditures occur before the end of the analysis period; that is, where the initial service life is less than the analysis period, and future maintenance, renovation, reconstruction, or rehabilitation (M,R&R) is needed. The equation for this situation, including user costs, is:

$$AC_{x1,n} = [\text{Equation 14.1}] + R_1 pwf_{i,a1} + R_2 pwf_{i,a2} + \ldots + R_j pwf_{i,aj}$$
$$- (1 - y/L)(R_1, R_2, \ldots, R_j) pwf_{i,\, a1,\, a2,\ldots aj} \qquad (14.4)$$

where $AC_{x1,n}$ = equivalent uniform annual cost for alternative $x1$, for an analysis period of n years;

$R_1, R_2, \ldots, R_j$ = costs of first, second, $\ldots$, jth M,R&R treatments;

$a_1, a_2, \ldots, a_j$ = ages at which the first, second, $\ldots$, jth M,R&R treatments occur, respectively;

y = number of years from time of last M,R&R treatment to end of analysis period; and

L = estimated life in years of last M,R&R treatment.

The present worth of benefits can be calculated in the same manner as the present worth of costs, using the following equation:

$$TPWB_{x1,n} = \sum_{t=0}^{n} pwf_{i,t}\left[(DUB)_{x1,t} + (IUB)_{x1,t} + (NUB)_{x1,t}\right] \qquad (14.5)$$

where $TPWB_{x1,n}$ = total present worth of benefits for alternative for an analysis period of n years;

$(DUB)_{x1,t}$ = direct user benefits accruing from alternative $x1$ in year t = indirect user benefits, if applicable, accruing from alternative $x1$ in year t;

$(IUB)_{x1,t}$ = indirect user benefits, if applicable, accruing from alternative $x1$ in year t; and

$(NUB)_{x1,t}$ = nonuser benefits, if applicable, accruing from alternative $x1$ in year t.

The net-present-value method follows from the foregoing methods—it is simply the difference between the present worth of benefits and the present worth of costs. Obviously, benefits must exceed costs if a project is to be justified on economic grounds. The equation for net present value is as follows:

$$NPV_{x1} = TPWB_{x1,n} - TPWC_{x1,n} \qquad (14.6)$$

where NPV_{x1} is the net present value of alternative $x1$; and $TPWB_{x1,n}$ and $TPWC_{x1,n}$ are as defined by eqs. (14.5) and (14.3), respectively.

However, for most infrastructure project alternatives, $x1$ in eq. (14.6) is not applicable directly to $x1$ itself, but rather to the difference between it and some other suitable alternative, say $x0$. Considering only direct user benefits, $(DUB)_{x1,t}$, these are then calculated as the user savings (resulting from lower operating costs, lower occupancy time costs, lower accident costs, etc.) realized by $x1$ over $x0$.

For some infrastructure situations, alternative $x0$ may be a base case of no capital expenditures (when increased maintenance and operating costs are required to keep the facility in service). The net-present-value method may then be expressed as:

$$NPV_{x1} = TPWC_{x0,n} - TPWC_{x1,n} \qquad (14.7)$$

where NPV_{x1} = net present value of alternative;

$TPWC_{x0,n}$ = total present worth of costs, for alternative (where $x0$ can be the standard or base alternative, or any feasible mutually exclusive alternative $x1, x2, \ldots xn$) for an analysis period of years; and

$TPWC_{x1,n}$ = as defined in eq. (14.3).

There are a number of advantages inherent in the present-worth method that make it particularly applicable to the infrastructure area compared to the "traditional" annual cost and benefit-cost methods. These advantages include the following:

1. The benefits and costs of a project are related and expressed as a single value.

2. Projects of different service lives, and with stage development, are directly and easily comparable.

3. All monetary costs and benefits are expressed in equivalent present-day terms.

4. Nonmonetary benefits (or costs) can be evaluated subjectively and handled with a cost-effectiveness evaluation.

5. The answer is expressed as a total payoff for the project.

6. The method is computationally simple and straightforward.

There are, however, several disadvantages to the present-worth method, including the following:

1. It cannot be applied to single alternatives when the benefits of those single alternatives cannot be estimated other than to simply calculate the present worth of costs.

2. The results, in terms of a lump sum, may not be as easily understandable to some people as a rate of return or annual cost. In some cases, the summation of costs in this form can act as a deterrent to investment because the lump sum seems so large compared to budgets.

14.7.3 Rate-of-return method

The rate-of-return method (sometimes called "internal rate of return") considers both costs and benefits, and determines the discount rate at which the costs and benefits, for a project are equal. It can be in terms of the rate at which the equivalent uniform annual cost is exactly equal to the equivalent uniform annual benefit, that is:

$$AC_{x1,n} = AB_{x1,n} \tag{14.8}$$

where $AC_{x1,n}$ = equivalent uniform annual cost for alternative $x1$ for an analysis period of years; and

$AB_{x1,n}$ = equivalent uniform annual benefit for alternative $x1$ for an analysis period in years.

Alternatively, the rate of return can be expressed in terms of the rate at which the present worth of costs is exactly equal to the present worth of benefits. This can be done by setting eqs. (14.3) and (14.5) equal to each other:

$$TPWC_{x1,n} = TPWB_{x1,n} \tag{14.9}$$

In applying the rate-of-return method, each alternative is first compared with the standard or base alternative to establish the difference in benefits. Then using eq. (14.9) or eq. (14.8), the rate of return can be calculated for alternatives $x1, x2, \ldots xn$. However, this is only a comparison with the standard. It is necessary to calculate the rate of return or comparison with the standard. It is also necessary to calculate the rate of return for comparisons between alternatives $x1, x2, \ldots xn$. This is done on the increase in costs between alternatives with successively higher first costs. Proceeding on the basis of such paired comparisons will eliminate all but one alternative—that one having the highest rate of return.

This alternative may or may not be economically attractive, depending on the decision for a minimum attractive rate of return. For example, if it were decided that an investment must have a minimum rate of return of 15 percent to be economically viable, any alternative yielding a lesser return would be rejected.

The rate-of-return method has a major advantage in that the results are well understood by most people. It is easy to comprehend a return on investment because of the association and familiarity with normal business terms. However, it must be remembered this is not a "real" interest rate in financial terms.

14.7.4 Benefit-cost-ratio method

The benefit-cost-ratio method has been widely used in the infrastructure area, particularly for large projects such as dams and causeways. It involves expressing the ratio of the present worth of benefits of an alternative to the present worth of costs, or the ratio of the equivalent uniform annual benefits to the equivalent uniform annual costs. The benefits are established by a comparison of alternatives. Using the present-worth formulation, which is preferred by most engineering economists, the benefit-cost ratio may be expressed as follows:

$$BCR_{xj,xk,n} = (TPWB_{xj} - TPWB_{xk})/(TPWC_{xj} - TPWC_{xk}) \qquad (14.10)$$

where $BCR_{xj,xk,n}$ = benefit-cost ratio of alternative xj compared to alternative xk (where xk yields the greater benefits, and represents the larger investment), over an analysis period of n years;

$TPWB_{xj,n}$, $TPWC_{xj,n}$ = total present worth of benefits, and of costs, respectively, for alternative xj; and

$TPWB_{xk,n}$, $TPWC_{xk,n}$ = total present worth of benefits, and of costs, respectively, for alternative xk.

The calculation of benefit-cost ratios for a set of proposed alternatives is first done by comparison with the standard or base alternative, using eq. (14.10). Alternatives exhibiting a ratio greater than 1.0 are compared on an incremental basis. This involves calculation of the benefit-cost ratio on the increments of expenditures for successively higher-cost alternative. Proceeding on the basis of such paired comparisons, using eq. (14.10), will reveal the alternative of highest economic desirability.

There are some disadvantages to the benefit-cost ratio method. A major drawback is the abstract nature of the ratio, which is difficult for some people to comprehend. Another disadvantage is the possible confusion over whether maintenance cost reductions should be in the numerator or the denominator, and whether cost reductions are "benefits" or "negative costs."

14.7.5 Cost-effectiveness method

The cost-effectiveness method can be used to compare alternatives if significant, appropriate measures of effectiveness can be established. This method is extensively used in the pavement field, where effectiveness of an alternative is calculated as the area under the performance curve (i.e., serviceability versus age) multiplied by traffic volume and length of road section. In this form it becomes a surrogate for vehicle operating costs, which are difficult to determine directly. It also becomes a convenient way to compare alternatives that have different performance curves.

Cost-effectiveness, then, is the ratio of effectiveness divided by the present worth of costs summarized over the life of the facility. It is the inverse of what is termed "cost-effectiveness," but it is used in this way because increasing ratios represent more attractive alternatives, as do increasing benefit-cost ratios. In the cost-effectiveness method, the actual numbers in the ratio are useful only in a comparative sense, because of the mixture of units. They are truly indexes of cost-effectiveness.

As a simple example, consider a facility where effectiveness is related to the number of service interruptions over time, as shown in Figure 14.1. Alternative X_1 involves an initial capital cost of $250,000 with a rehabilitation cost of $80,000 at year seven. It is estimated to have a residual value of $25,000 at the end of the analysis period. Alternative X_2 requires an initial capital expenditure of $400,000. It is estimated to have a residual value of $65,000 at the end of the analysis period. The floor areas of X_1 and X_2 are 165 m² and 265 m², respectively. Assume that maintenance costs are relatively small for each alternative, and that the traffic processed by each is approximately the same.

Calculation of effectiveness, E, for these two alternatives can then be carried out as follows:

$$E = (\text{area under curve}) \times (\text{floor area}) \qquad (14.11a)$$

$$E_{x_1} = 442 \times 165 = 72,930 \qquad (14.11b)$$

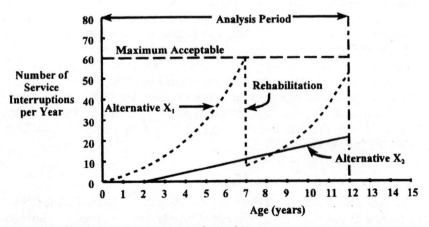

Figure 14.1   Level of service versus age relationship.

$$E_{x_2} = 620 \times 265 = 164{,}300 \qquad (14.11c)$$

Present worth of costs, C, for the alternatives using a discount rate of 4 percent and analysis period of 12 years, is calculated as follows:

$$C_{x_1} = \text{initial cost} + (\text{rehabilitation cost at year 7}) \times (PWF_{y=7})$$
$$- (\text{residual value at year 12}) \times (PWF_{y=12}) \qquad (14.12a)$$

$$C_{x_1} = 250{,}000 + 80{,}000 \times \frac{1}{(1 + 0.04)^7} - 25{,}000 \times \frac{1}{(1 + 0.04)^{12}} = 295{,}178$$
$$(14.12b)$$

$$C_{x_2} = 400{,}000 - 65{,}000 \times \frac{1}{(1 + 0.04)^{12}} = 359{,}401 \qquad (14.12c)$$

Cost-effectiveness ratios, E/C, would be equal to:

$$E_{x_1}/C_{x_1} = 72{,}930/295{,}178 = 0.25 \qquad (14.13a)$$

$$E_{x_2}/C_{x_2} = 164{,}300/359{,}401 = 0.46 \qquad (14.13b)$$

On a comparative basis, over the analysis period or life-cycle, alternative X_2 is the more attractive alternative, even though it has a significantly higher cost.

14.8 Selecting an Appropriate Economic Analysis Method

There are several basic considerations in selecting the most appropriate method for economic evaluation of alternatives. They include the following factors:

- How important is the initial capital expenditure in comparison to future expected expenditures? Often, public officials and private interests are concerned primarily with initial costs. An economic analysis may indicate, for example, that a low capital expenditure today can result in excessive future costs for a particular alternative. Yet the low capital expenditure may be the most relevant consideration to decision makers, especially in the face of limited funding.

- What method of analysis is most understandable to the decision maker? The method used may not be technically the best for the situation at hand, but it provides the necessary decision support and understanding, and thus has a high likelihood of being used.

- What method suits the requirements of the particular agency involved? Although the present-worth method may be preferable for many public agencies, an annual cost method might be more suitable for a privately owned facility.

- Are measures of benefits or effectiveness to be included in the analysis? Any method that does not consider the differences in benefits or effectiveness between alternatives is basically incomplete for use by a public agency. However, for a private-sector situation, an implicit assumption of equal benefits for various alternatives may be satisfactory.

14.9 Effect of Discount Rate

It was noted in Sec. 14.5 that the discount rate selected by most agencies is a policy decision, but usually it is the difference between the interest rate for borrowing money and the inflation rate. In the example given in Sec. 14.7.5, a discount rate of 4 percent was used.

Often, however, it is desired to assess the sensitivity of the economic analysis result to discount rates higher and lower than the discount rate calculated. A convenient way to illustrate the effect of such discount-rate changes is provided in Figure 14.2 [AISI 90]. It represents

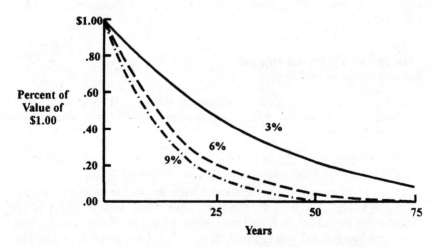

Year	Discount Rate		
	3%	6%	9%
0	1.00	1.00	1.00
25	.48	.23	.12
50	.23	.05	.01
75	.11	.01	.01

Figure 14.2 Present value of $1 expended at various intervals and discount rates. [AISI 90].

the use of eq. (14.2) to show the present worth of $1 expended at some-time in the future for three discount rates. The effect shown has greater significance to future spending at a low discount rate, and less significance at a high discount rate. For example, $1 expended 25 years in the future has a NPV of 48¢ at 3 percent, 23¢ at 6 percent, and only 12¢ at 9 percent. Note that when expended at 75 years, the NPV is nil for both 6 percent and 9 percent, and about 11¢ at 3 percent.

14.10 Application to Highways

Haas, Hudson, and Zaniewski [Haas 94] present detailed examples of life-cycle analysis for road and highway pavements. Vehicle operating cost is an important component of road-user cost, which depends upon the vehicle type, speed, pavement geometry and grade, and pavement condition. Reduction in vehicle operating costs due to improved pavement condition and/or reduced congestion can be considered as a user benefit. The Federal Highway Administration and the World Bank have developed vehicle-operating-cost models applicable to pavement life-cycle analysis [Zaniewski 82, Watanatada 87, Uddin 87, Uddin 93, Paterson 92].

Life-cycle cost analysis is also recognized for bridge management system development [Hudson 87]. A NIST (National Institute of Standards and Technology) report [CDA 96] outlines the application of life-cycle cost analysis using new technology materials, such as the use of fiber-reinforced plastics (FRP) to retrofit and repair bridges in seismic areas. According to this report, the minimum NPV of a life-cycle cost method can be used to satisfy the Intermodal Surface Transportation Efficiency Act's requirement that life-cycle costs be considered in the design of transportation-related structures, and Executive Order 12893, which requires that the costs of federal infrastructure investment be accounted for over the life span of each project [CDA 96].

14.11 Application to Water and Sewer Mains

Economic analysis to compare repair versus replacement M,R&R alternatives for water and sewer mains has been used by agencies in many cities, such as New York City, Dallas, Texas, Washington, D.C., and Seattle, Washington [O'Day 84, Steinthal 84, Thomasson 82, Urban 84c]. In general, replacement alternatives are more favorable when mains have deteriorated structurally and shown a consistent record of frequent breaks. Table 14.1 shows the results of Dallas water-main economic analysis [after Steinthal 84] to calculate the least life-cycle cost alternative based on the previous breaks record and a 20-year analysis period.

TABLE 14.1   Economic Analysis for Dallas Water Main Alternatives [Steinthal 84]

Main size	Year break										
	71	72	73	74	75	76	77	78	79	80	81
15 cm (6 in)		/	/					/	///	//// ////	///

Breaks per year in the last four years = 16/4.   Length of main = 250 ft.   Breaks per year per 1000 ft = (16/4)/(250/1000) = 16.

Cumulative repair cost = 360.34 × 16/4 × 75.477 = 108,789.53
                          ($/break) (Breaks/year) (CAF)[1]   ($)

Cumulative replacement cost = 20.5 × 250 × 0.1095 × 20 = 11,223.75
                              ($/ft)   (ft)   (CRF)[2]   (Years)   ($)

Analysis conducted for 20-year cost of both repairing and replacing a given water main whenever the water-main break rate exceeds 1.8 breaks per year per 1000 feet of main (over the last four years). The above information given at 12.4 percent interest and 9 percent bonds.

(CAF)[1] = Capital amount factor and (CRF)[2] = Capital recovery factor.

Risk analysis, including lost revenues, cost of lost time, and inconvenience for each alternative, should also be included in any analysis that considers the reduction in the user costs as "benefits." A substantial leakage-control program can provide considerably more benefits than the costs involved in the identification and repair of leaking segments, as noted by O'Day. The following example is from a major United States water-utility study conducted in 1981, which resulted in the location and repair of 596 leaks, and the elimination of 12.2 million gallons of wasted water per day in the 200 mi of the water-distribution system [O'Day 84]:

> The leak detection costs for the utility staff, consultant fees, and equipment costs were approximately $204,000, equivalent to slightly more than $1000 per mile for the 200 miles in the survey area. The total costs for repair were $243,000 for a total detection and repair cost of $447,000.
> The value of the future water losses saved by this leak control program is equivalent to the present value of future water and waste-water treatment and pumping costs. The net present worth for these savings is estimated to be nearly $6,600,000 over a 30-year period. That is, the $447,000 leak detection and repair program will save an estimated $6,000,000 in treatment and pumping costs over a 20-year period.

The foregoing case example shows that the benefits in terms of savings are 14.42 times the costs. For a cost of $2,000 per mile, significant water loss was saved, future causes of main breaks were

TABLE 14.2   Life-cycle Cost Analysis for Sewer-main Material Selection [AISI 90]

Data and analysis method	Alternative A	Alternative B	Alternative C
Description	Galvanized Corrugated steel pipe (CSP)*	Asphalt-coated CSP	Reinforced concrete pipe (RCP)
Initial costs	$195,000	$214,500	$230,000
Other costs	$48,750*	—	—
Design life	50 years	50 years	50 years
Discount rate	9%	9%	9%
Current cost	$195,000 + 48,750 Total = $243,750	$214,500 + 0 Total = $214,500	$230,000 + 0 Total = $230,000
Present-worth method [Present value (at 9%)]	$195,000 + (48,750 × 0.318 = $1,550) = ($196,550 (lowest)	$214,500	$230,000
Equivalent uniform annual cost (50 years at 9%)	$17,930 (lowest)	$19,560	$20,980

| Differential cash-flow method | Compare alternative A and C | | |
	Alternative A	Alternative C	Difference (C−A)
Year 0	$195,000	$230,000	$35,000
Year 40	$48,750	—	(48,750)
Total	$243,750	$230,000	$(13,750)

(Rate of return, %)	To avoid future expenditure of $48,750 in Alternative A, the added initial cost of $35,000 in Alternative C results in 0.83%, or less than 1% internal rate of return. This a poor rate of return on the initial investment.
Effect of salvage	Not significant; at even 30% of the original cost, present worth of salvage represents less than 5% of the initial cost.

*Alternative A requires invert maintenance at 25% of initial cost due to a projected life of 40 years for the galvanized CSP ($48,750).

reduced, and only less than 1 mi out of 200 mi of water mains were replaced [O'Day 84]. An example of life-cycle cost analysis for sewer-main material-type selection is presented in Table 14.2 [AISI 90].

14.12   Application to Buildings

Prediction of effective service life is more complex for buildings, because the structural integrity of a building depends upon many factors aside from the materials of construction and the performance of the various functional subsystems. These complexities generally have discouraged the use of routine life-cycle cost analysis for design and

M,R&R programming of buildings in the past. In *Pay Now or Pay Later* [BRB 91], the Building Research Board provides general guidelines for life-cycle cost analysis and gives an example in the Department of Veterans Affairs (VA), which owns and operates the United States' largest system of hospitals (172 medical centers, some 2,000 buildings). The VA has used a 40-year analysis period and a 5 percent discount rate for their life-cycle cost analysis [BRB 91]. The study highlighted seven factors that could reduce the cost of ownership:

1. Reduced tie to design and construction
2. Standardization of building components and utility system concentration
3. Improved space utilization and layout of mechanical system
4. More cost-effective maintenance and minor alternations, using materials with better performance and access for service
5. Optimization of heating, ventilation, and air conditioning
6. Better design features to improve housekeeping
7. More cost-effective rehabilitation alternatives

The total life-cycle cost of a hospital was 8.7 times the initial cost; however, the present worth of savings was estimated to be approximately 10 percent of the life-cycle cost.

The National Institute of Standards and Technology (NIST) has also developed a computer program, *Building Life-Cycle Cost (BLCC)*, and is currently involved in an international effort to standardize service life-cycle analysis procedures [AIJ 93, BS 92, CSA 94, Frohnsdorff 96, NIST 95]. The Building Maintenance, Repair, and Replacement Database (BMDB) computer program offered by the American Society for Testing and Materials (ASTM) can be used to estimate data for building life-cycle cost analysis [ASTM 90]. The BMDB program generates maintenance costs by applying labor and equipment rates, material-location adjustment factors, cost-escalation factors, and discount rate. The material costs (in 1986 dollars) are based on Washington, D.C., and are adjusted to other locations through the application of a geographical location adjustment factor. Table 14.3 shows an example of data generated from the BMDB program for life-cycle cost analysis [ASTM 90].

14.13 Summary

Economic analysis is an essential part of infrastructure management. The methods and applications may vary, as shown here and in the references; they may not, however, be omitted, if good results are

TABLE 14.3   An Example of Data Generated from the BMDB Program for Life-Cycle Cost Analysis [after ASTM 90]

Resource and Unit Cost—Present Value of All 25 Years of Maintenance and Repair Costs ($ per Unit Measure)

(a) Component description	Unit measure	Annual maintenance and repair			Unit cost ($)
		Labor	Material	Equipment	
Architecture					
Roofing					
Roof covering					
Built-up roofing	SF	0.00488	0.03171	0.00244	0.14
Place new membrane over existing					
Modified Bitumen/Thermoplastic	SF	0.00248	0.03218	0.00119	0.09
Membrane Replacement or Repair					
Thermosetting	SF	0.00174	0.02202	0.00088	0.06
Membrane replacement— thermosetting					
Slate	SF	0.00259	0.01458	0.00124	0.07
Cement asbestos	SF	0.00254	0.03402	0.00122	0.09
Tile	SF	0.00217	0.02934	0.00103	0.08
Roll roofing	SF	0.00754	0.01556	0.00386	0.18
Total roof replacement-roll roof					
Shingles	SF	0.00259	0.02384	0.00140	0.08
Replace new over existing shingle					
Metal	SF	0.00204	0.01546	0.00103	0.06
Fiberglass rigid stp. roof	SF	0.00232	0.06266	0.00113	0.11
Concrete, sealed panel roof	SF	0.00601	0.01643	0.00297	0.15
Concrete, sealed panel RF4	SF	0.00544	0.01175	0.00282	0.13
Concrete sealed poured	SF	0.01374	0.08807	0.00692	0.39
Fiberglass, rigid roof	SF	0.00463	0.06266	0.00236	0.16

expected from IMS. Many agencies have economic analysis capabilities in house, but do not use them effectively. The methods are relatively simple and straightforward, and should be coded directly into any IMS software one adopts.

TABLE 14.3   An Example of Data Generated from the BMDB Program for Life-Cycle Cost Analysis [ASTM 90] (*Continued*)

Resource and Unit Cost—Present Value of All 25 Years of Maintenance and Repair Costs ($ per Unit Measure)

			Replacement and High-Cost Tasks			
(b) Component description	Unit measure	Year	Labor	Material	Equipment	Unit cost ($)
Architecture						
Roofing						
Roof covering						
Built-up roofing	SF	28	0.04938	0.70490	0.02469	1.79
Place new membrane over existing	SF	14	0.02414	0.69960	0.01207	1.23
Mod. bit./thermoplastic Membrane replacement or repair	SF	20	0.05659	0.85860	0.02829	2.11
Thermosetting	SF	20	0.05659	0.85860	0.02829	2.11
	SF	20	0.03683	0.69960	0.01841	1.51
Membrane replacement —thermo	SF	20	0.03683	0.69960	0.01841	1.51
Slate	SF	70	0.06885	6.04200	0.03442	7.56
Cement asbestos	SF	70	0.05437	0.75190	0.02718	1.95
Tile	SF	70	0.10169	3.07400	0.05084	5.32
Roll roofing	SF	10	0.04141	0.74963	0.02070	1.66
Total roof replacement —Roll roof	SF	10	0.04141	0.74963	0.02070	1.66
Shingles	SF	40	0.04118	0.74497	0.02059	1.65
Replace new over existing shingle	SF	20	0.02996	0.43460	0.01498	1.10
Metal	SF	30	0.36265	2.17300	0.18132	10.17
Fiberglas rigid stp. roof	SF	20	0.04543	6.01550	0.02272	7.02
Concrete, sealed panel roof	SF	60	0.06123	24.07419	0.03061	25.42
Concrete, sealed panel RF4	SF	300	0.04342	24.07419	0.02171	25.03
Concrete sealed poured	SF	500	3.81056	18.03219	1.90528	102.0
Fiberglass, rigid roof	SF	20	0.04133	6.01550	0.02066	6.93
Total roof replacement —fiberglass	SF	20	0.04133	6.01550	0.02066	6.93

Location: Washington, D.C. Study starts three years before beneficial use. Discount rate = 10%.

References

[AIJ 93]   Architectural Institute of Japan, *The English Edition of Principal Guide for Service Life Planning of Buildings,* Japan, 1993.

[AISI 90]   *Modern Sewer Design,* American Iron and Steel Institute (AISI), Washington, D.C., 1990.

[Army 86]   "Economic Studies for Military Construction Design-Applications," *TM 5-802-1,* Department of the Army, Washington, D.C., 1986.

[ASTM 90]   ASTM, "Building Maintenance, Repair, and Replacement Database (BMDB) for Life-Cycle Cost Analysis," *A User's Guide to the Computer Program,* American Society for Testing and Materials, Philadelphia, Pa., 1990.

[BRB 91]   Building Research Board, *Pay Now or Pay Later,* National Research Council, Washington, D.C., 1991.

[BS 92]   *Guide to Durability of Buildings and Building Elements, Products and Components,* BS 7543:1992, British Standards Institution, London, U.K., 1992.

[CDA 96]   "NIST Develops Methodology to Evaluate Life-Cycle Costs of Composites in Bridge Applications," *Composites Design and Applications,* March 1995, p. 6.

[CSA 94]   *Guideline on Durability in Buildings,* CSA S478, Draft 9, Canadian Standards Association, Sept. 1994.

[DOI 83]   "Economic and Environmental Principles and Guidelines for Water and Related Land Resources Implementation Studies," *Report PB 84-199405,* U.S. Water Resources Council, Department of the Interior, Washington, D.C., 1983.

[Frohnsdorff 96]   G. Frohnsdorff, "Predicting the Service Lives of Materials of Construction," *Proceedings, Fourth Materials Engineering Congress,* Washington, D.C., Nov. 1996, Vol. 1, pp. 38–53.

[Haas 94]   R. Haas, W. R. Hudson, and J. P. Zaniewski, *Modern Pavement Management,* Krieger Publishing Company, Malabar, Fla., 1994.

[Hudson 87]   S. W. Hudson, R. F. Carmichael III, L. O. Moser, W. R. Hudson, and W. J. Wilkes, "Bridge Management Systems," *NCHRP Report 300,* National Cooperative Highway Research Program, Transportation Research Board, National Research Council, Washington, D.C., 1987.

[NIST 95]   Personal communications of W. Uddin with G. Frohnsdorff, Chief of Building Materials Division, National Institute of Standards and Technology, 1995.

[O'Day 84]   D. K. O'Day, "Aging Water Supply Systems: Repair or Replace," *Infrastructure—Maintenance and Repair of Public Works, Annals of the New York Academy of Science,* Vol. 431, Dec. 1984, pp. 241–258.

[Paterson 92]   W. D. O. Paterson and B. Attoh-Okine, "Simplified Models of Paved Road Deterioration based on HDM-III," in *Transportation Research Record 1344,* Transportation Research Board, Washington, D.C., 1992.

[Steinthal 84]   B. G. Steinthal, "Infrastructure Maintenance Strategies—Governmental Choices and Decision Methodologies," *Infrastructure—Maintenance and Repair of Public Works, Annals of the New York Academy of Sciences,* Vol. 431, Dec. 1984, pp. 139–154.

[Thomasson 82]   R. O. Thomasson, "In-Place Sewer Reconstruction Proves Cost-Effective," *American City and County,* Feb. 1982.

[TRB 85]   Transportation Research Board, "Life-Cycle Cost Analysis of Pavements," *Synthesis of Highway Practice No. 122,* National Cooperative Highway Research Program, National Research Council, Washington, D.C., 1985.

[Uddin 93]   W. Uddin, "Application of User Cost and Benefit Analysis for Pavement Management and Transportation Planning," *Proceedings, 4R Conference and Road Show,* Philadelphia, Pa., Dec. 1993, pp. 24–27.

[Uddin 87]   W. Uddin, R. F. Carmichael III, and W. R. Hudson, "A Methodology for Life-Cycle Cost Analysis of Pavements Using Microcomputer," *Proceedings, 6th International Conference, Structural Design of Asphalt Pavements,* Ann Arbor, Mich., Vol. 1, 1987.

[Urban 84c]   H. P. Hatry and B. G. Steinthal, *Guides to Managing Urban Capital Series,* Vol. 4: *Guide to Selecting Maintenance Strategies for Capital Facilities,* The Urban Institute, Washington, D.C., 1984.

[Watanatada 87]   Watantada, Harral, Paterson, Bhandan, and Tsunokawa, *The Highway Design and Maintenance Standards Model,* Vols. 1 and 2, Johns Hopkins University Press, Baltimore, Md., 1987.

[Winfrey 69]   R. Winfrey, *Economic Analysis for Highways,* International Textbook, 1969.

[Zaniewski 82]   J. P. Zaniewski, B. C. Butler, G. Cunningham, G. E. Elkins, M. S. Paggi, and R. Machemehl, "Vehicle Operating Costs, Field Consumption and Pavement Type and Condition Factors," *Final Report,* Federal Highway Administration, Washington, D.C., 1982.

15

Prioritization, Optimization, and Work Programs

15.1 Introduction

It was shown in Chapter 8 that when the serviceability or quality of an infrastructure component reaches an unacceptable or intervention level, some action is needed. This level could, for example, be in terms of the maximum allowable number of annual service interruptions shown in Figure 14.1, the maximum allowable user-delay time on a facility, minimum flow capacity in a pipe network, and so forth. In other words, the component or facility reached a condition where action is needed. Such action might consist of M,R&R, or simply increased routine maintenance to keep the facility from exceeding the intervention level.

If a facility is currently at the intervention level, it is a "now need." Future needs would be determined on the basis of the performance or deterioration prediction model (i.e., the time or year that the facility or component is estimated to reach the intervention level, which was, for example, seven years for alternative X_1, in Figure 14.1).

If sufficient funds are available (i.e., an unconstrained budget), all needs can be addressed when they occur. The usual situation, however, for most public infrastructure is a limited or constrained budget. In such cases, priorities have to be set on what action or work will be undertaken, where, and when. This chapter is directed to the procedures that can be used to establish priorities, the role of optimization in priority analysis, budgeting issues, and the development of work programs under constrained budgets.

15.2   Framework for Prioritization

Prioritization as a whole involves four central steps: (1) information acquisition; (2) processing and interpretation; (3) determination of current needs and future needs; and (4) priority analysis and results, as shown in Figure 15.1. Additionally, there are key associated elements of deterioration or performance models, intervention levels, budget constraints, program period selected, and available alternative strategies.

The actual models, methods, and procedures within the framework of Figure 15.1 can range from very simple to very sophisticated. For example, the prioritization might simply involve some basic inventory data, an annual (year-by-year) program period, a determination of needs, a priority analysis, and a recommended program of work selected by engineering judgment. At the other extreme it may involve a comprehensive database, a multi-year program period, multifactor deterioration models, decision analysis procedures to identify feasible alternative strategies, a mathematical programming method for priority analysis, and the establishment of work programs based on optimization.

Whatever approach is used, the final result, in terms of the recommended progress of work, should answer the following questions:

1. Which projects or infrastructure components should receive action?

2. What action (M,R&R) should be applied?

3. When should the work be done?

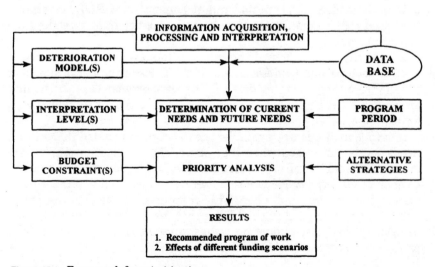

Figure 15.1 Framework for prioritization.

A methodology, within the framework of Figure 15.1, that can deal with these questions concurrently will not be simple. However, if a truly optimal expenditure of available funds or budget is to be achieved, all possible combinations of which, what, and when should be considered and evaluated.

15.2.1 Selecting a program period

The length of the program period for M,R&R improvements is not necessarily the same as that for life-cycle economic evaluation. In fact, it would usually be less, because the latter commonly covers 20 or more years. Certainly, the length of the program period for maintenance would generally only be one to five years.

The question then becomes one of what is both a reasonable and useful program period for rehabilitation. It can range from single-year-by-single-year to multi-year programs. The former is convenient, but cannot consider alternative action years for rehabilitation. Thus, timing is excluded from the optimization.

The following guidelines for rehabilitation program periods are considered reasonable and have found substantial acceptance in practice:

1. Five years for infrastructure networks or portions of networks (roads, pipe distribution systems, and so forth), where there is considerable uncertainty for future funding and/or no useful purpose is seen in developing even tentative priority programs beyond that time.

2. Ten years for other network situations, particularly when it is desired to evaluate the long-term, strategic implications of greater or lesser levels of funding.

A very practical approach is to develop a 5- or 10-year program. However, considering that uncertainties increase with time, it is more practical to fix only the first two or three years of the program, and carry out annual or biannual updates. In this way, the flexibility exists for some projects to be advanced one or two years if costs or bid prices come in below estimates. Alternatively, if costs are higher than expected, then some first-year program projects may be deferred to the second or later years.

15.3 Priority Analysis Methods

Priorities can be determined by many methods, ranging from simple subjective ranking to a true optimization in which all possible combinations of "what project, when, and where" are evaluated in terms of an output parameter or objective function, as subsequently discussed.

TABLE 15.1   Classes of Methods that Can Be Used in Priority Analysis

Class of method	Advantages/disadvantages
Simple subjective ranking of projects based on judgment	Quick, simple; subject to bias and inconsistency; may be far from optimal
Ranking based on parameters, such as level of service and condition	Simple and easy to use; may be far from optimal
Ranking based on parameters with economic analysis	Reasonably simple; should be closer to optimal
Optimization by mathematical programming model for year-by-year basis	Less simple; may be close to optimal; effects of timing not considered
Near optimization using a marginal cost-effectiveness approach	Reasonably simple; close to optimal results
Comprehensive optimization by mathematical programming model, taking into account the effects of "which, what, and when"	Most complex; can give optimal program (maximization of benefits, minimization of costs)

Table 15.1 summarizes the various classes of methods, plus their advantages and disadvantages.

Simple subjective ranking is easy to apply in most situations, but the results may not be the best expenditure of funds or very defensible. Moreover, its use in other than very straightforward cases (options are few and priorities are relatively simple to determine), obviously defeats the principles of an infrastructure management system. However, rankings can be weighted, for example, by using traffic, volume of flows, and amounts of usage, depending on the infrastructure involved, and a much more rational and defensible result can be achieved. Such results may still be less than optimal, especially if a significant number of possible alternative treatment and timing combinations exist for each project.

If the ranking also includes an economic analysis, a closer-to-optimal result is more likely to be ensured. Optimization on a year-by-year basis may or may not achieve an overall closer-to-optimal result than the foregoing method, but has a higher likelihood of doing so.

15.3.1   Priority setting based on ranking

Capital budgeting decisions at the network level are commonly based on ranking or rating criteria. The ratings are combined in an overall priority score using appropriate weights to adjust for importance of a project and safety. For example, the City of Dallas categorizes storm-water improvement projects based on: (1) possible loss of life; and (2) flood water in residential or commercial properties, or in surrounding areas to a depth that disrupts emergency-vehicle access [Steinthal 84].

Initial prioritization of water-main projects in Dallas and Milwaukee used ratings based on "breaks per year per 333 m (1,000 ft) of pipe" and "breaks per 33 m (100 ft)," respectively. Large water-main projects in Seattle are selected based on a composite overall rating considering the following factors [Steinthal 84]:

- Visual inspection of lining condition (0–10 points)
- Effect of the lining on water quality (1–10 points)
- Degree of pitting in the pipe (0–10 points)
- Leak listing over ten years (0–5 points)
- Necessity of the water main for area supply (0–5 points)

Hatry and Steinthal [Urban 84c] reviewed the priority-setting procedures of many public works agencies, and presented the following summary of rating criteria that are used by others:

- Need
  1. Condition/structural integrity
  2. Frequency/extent of existing problems
  3. Expected lifetime of existing infrastructure
  4. Likelihood of future problems/danger
  5. Risk/magnitude of future problems
  6. Consequences of delays
  7. Customer complaints
  8. Availability of emergency backup

- Environmental Factors
  1. Geographic disbursement/equity
  2. Probable support/opposition
  3. User costs and preferences
  4. Community disruption
  5. Areawide needs
  6. Number of users benefited
  7. Importance to transit or commerce

- Resources
  1. Funding availability/capital costs
  2. Operating and maintenance costs
  3. Staff workloads
  4. Probability of completion

15.3.2 Near-optimization methods

Near optimization, based on a marginal cost-effectiveness method, has found considerable application in the roads-and-pavements area [Haas 94], and could also be applied in many other infrastructure areas. It can provide essentially the same results as a true optimization, and may be summarized in terms of the following steps:

1. Consider each combination of section (component), treatment alternative, and year in the program period (i.e., the "which, what, and when" combination)

2. Calculate the effectiveness (E) of each combination (that is, the area under the performance curve, weighted by usage, length, and area. See Sec. 14.7.5).

3. Calculate the cost (C) in present worth terms, of each treatment alternative in each combination.

4. Calculate the cost-effectiveness (CE) of each combination as the ratio of effectiveness divided by costs.

5. Select the combination of treatment alternative and year for each section that has the best CE, until the budget is exhausted.

6. Calculate the marginal cost-effectiveness (MCE) of all other strategies (i.e., "which, what, and when") for each section, as follows:

$$MCE = (E_s - E_r)/(C_s - C_r) \qquad (15.1)$$

where E_s = effectiveness of the combination selected in Step 5;
E_r = effectiveness of the combination for comparison;
C_s = cost of the combination selected in Step 5; and
C_r = cost of the combination for comparison.

7. If *MCE* is negative, or if $E_r < E_s$, then the comparative strategy is eliminated from further consideration; if not, it replaces the combination selected in Step 5.

8. The process is repeated until no further selections can be made in any year of the program period (i.e., when the budget is exhausted).

15.3.3 Prioritization based on optimization

Optimization based on mathematical programming models can be used for either single-year or multiyear prioritization. The results do not indicate priorities per se; rather, alternatives are selected to satisfy a specific objective function (i.e., a value is maximized, such as effectiveness, or minimized, such as costs). Formulations used in these models include several variations of linear programming and dynamic programming. Linear programming (LP) is a useful formulation for multi-year prioritization because it can be set up to model the trade-offs between project timing and benefit losses. Each possible implementation, or action, year can be treated as an independent alternative, along with the within-project alternatives and the projects or sections in the network. Thus, all possible combinations are considered and compared.

The formulation of an LP model for maximizing the total present value of m projects, each with k within-project alternatives, for a T-year programming period can be done in the following way:

$$\text{Maximize} \sum_{i=1}^{m} \sum_{j=1}^{k} \sum_{t=1}^{T} X_{ijt} \times B_{ijt} \tag{15.2}$$

$$\text{Subject to} \sum_{t=1}^{T} \sum_{j=1}^{k} X_{ijt} \leq 1 \text{ for } i = 1, 2, \dots m \tag{15.3}$$

$$\sum_{t=1}^{T} \sum_{j=1}^{k} X_{ijt} \times D_{ijtt''} \leq B_t \text{ for } t = 1, 2, \dots T \tag{15.4}$$

and

$$X_{ijt} \leq 0 \tag{15.5}$$

where X_{ijt} = section i (of m total sections) with alternative j (of k total treatment alternatives) in year t (of the T years in the program period);

B_{ijt} = present value of annual benefits (including salvage value) of section i, with alternative j, built in year t, all discounted to base year at a discount rate of R;

$D_{ijtt''}$ = the actual construction and/or maintenance cost of section i, with alternative j, built in year t, incurred in years t''; and

B_t = budget for year t.

X_{ijt} is a discrete variable equal to 1 if a project is selected for section i in year t, and equal to zero otherwise.

Equation (15.2) is the objective function for maximization of benefits. Maximization of net benefits (benefits minus costs) is not necessary, because construction and maintenance costs are specifically dealt with in the annual budget constraints. Thus, there is no need to discount costs back to a base year to calculate the net present value of the projects.

Equation (15.3) states that an X_{ijt} combination is unique; that is, a section cannot be selected twice during the analysis period. Either it is, or is not, built. If a project is committed (by judgment, political intervention, and so forth), then the constraint becomes one of equality (i.e., $X_{ijt} = 1$).

Equation (15.4) represents the budget constraint each year in the programming period, which cannot be exceeded. It states that the summation over m total sections and k total alternative treatments of all the X_{ijt} costs must be less than or equal to the budget, B_t.

Equation (15.5), a common constraint of all linear programming problems, states that it is not possible to recapture construction costs by not constructing the project.

It should be noted that costs are considered directly in the LP model up to the end of the programming period. After that, costs cannot be taken into account because of the absence of budget constraints. It is for this reason that costs incurred in each year after the programming period are considered in the analysis indirectly by subtracting them from the respective annual benefits.

15.4 Budgeting and Financial Planning Issues

Several possible approaches exist for setting infrastructure expenditure budgets. They include the following:

1. Extrapolation from previous years and adjusting for additions or growth to the system (i.e., the water-distribution network has grown by 5 percent and for inflation).

2. Demand based on condition-responsive budgeting (see Sec. 13.4), which programs work to respond to expected performance indicators or conditions.

3. Budgeting based on a desired level of performance or level of service (which then becomes a financial-planning tool).

4. Budgeting based on achieving an optimum level of service for the network or system of assets as a whole.

5. Budgeting based on affordability, as determined by agency planners or administrators, or by elected bodies.

A key issue for all of the above approaches is whether the budgets are split or separated into different categories for expenditures of: (1) maintenance and rehabilitation; (2) new construction; and (3) reconstruction. Traditionally many larger agencies, and in particular state transportation departments, have operated by creating separate expenditure categories for maintenance and capital spending. This often seems less a matter of rational policy than of organizational structure (i.e., maintenance division and planning division might each be responsible for its own programs). In the case of the privatized road network described in Sec. 13.3, this is not an issue, since a single entity (the contractor) is responsible for all the work and is able to optimize the budget.

The first approach, setting a budget by extrapolation and minor adjustments, has been common in road, water, sewer, parks, buildings, and various other infrastructure components maintenance. It is relatively simple, and uses the experience from previous years. But it also represents business as usual, may well involve a significantly less than optimal use of funds, and may tend to discourage innovation.

The second approach, demand-based or condition-responsive budgeting, particularly in the case of maintenance, is quite rational because it should involve only necessary expenditures. However, it can also result in actual expenditures that vary greatly from the budget. For example, severe, unexpected circumstances, such as an extreme winter or flooding, could result in a high number of breaks or interruptions of service requiring extra maintenance.

Budgeting based on a desired level of performance, the third approach, can also be quite rational, and can result in a high degree of system stability. For example, consider a city's network of recreational paths, which are assessed in terms of a composite condition index on a scale of 0 to 10. Figure 15.2 shows the variation of the average condition index of the network for four budget levels ranging from $0 per year to $200,000 per year. The current average condition index is approximately 5, and would stay at that level for budget level 2 ($100,000 per year). If that average condition index represents the

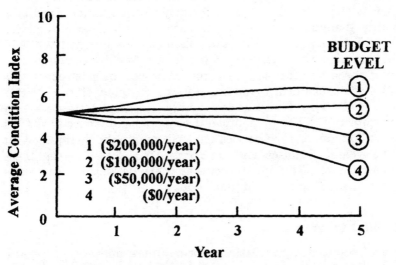

Figure 15.2 Variation of the average condition index of a city's recreational path network with different annual budget levels.

minimum desired level of performance, then at least $100,000 per year needs to be spent on M,R&R.

The fourth approach, budgets set to optimize some factor, such as benefits, is perhaps the most rational, but may run up against the issue of affordability. In this approach, expenditures would be at a level where the marginal benefits achieved approach zero; in other words, increasing the budget would not bring any increased benefits and would represent "gold plating." A good example would be a road network where vehicle-operating-cost savings have increased thanks to increased expenditures for smoother pavements or greater capacity. At some point, however, the vehicle-operating costs level out, and, additional expenditures would represent gold plating and would not be warranted.

A budgeting approach based on affordability seems to have become quite popular in the case of roads, hospitals, and various other public assets. However, such an approach can result in serious problems and a decline in asset value if the affordability is set at too low a level. Such an approach only works at all in the case of reasonably affluent agencies.

15.5 Budget Allocation Issues

Budget allocation can become a particularly difficult issue when different functional classes and/or different degrees of traffic/usage are involved in the same analysis. For example, on a purely economic basis (i.e., to maximize benefits) a state transport authority would expend its budget on freeways in the major urban centers, and allocate little, if anything, to low-volume rural roads because of the large traffic differential. As another example, hospitals in small towns or remote locations with relatively few patients would receive minimal funding based on total service alone. However, in most cases this would be quite politically and socially unacceptable, since everyone deserves some level of hospital availability.

Consequently, policy or political decisions usually specify that at least a minimum level of service and accessibility be maintained for such low-volume facilities, and some budget is allocated to meet these minimum targets. This illustrates that policy decisions are interwoven with levels of economic optimization.

15.6 Work Programs

The final result of a prioritization, or optimization, process, regardless of type, should be a recommended work program. This is usually in terms of a listing of sections, components, or units; their location;

the type of work or action to be undertaken on each (i.e., pipe relining and road-surface treatment); and the estimated cost. Usually, this is done on a yearly basis. The next operational step would be to establish a work schedule, allocation of resources (people, materials, and equipment), and quality control/assurance planning in accordance with the work program.

It should be noted that any work program is only as good as the data, models, and assumptions used in developing it. Frequently, data is collected well in advance of the program development, and conditions may have changed between the time the data was collected and the program is formulated. In addition, the priority, or optimization, model may not capture all of the factors that need to be considered for each individual section in a network. In large organizations, the programming is often performed by the central office. Hence, there may be a need for adjustments by the local engineer to assure the applicability of the program. The infrastructure management system and the prioritization activity provides decision support; it does not make the decisions. These are always reserved for the decision-maker or governing body.

References

[Haas 94]   R. Haas, W. R. Hudson, and J. Zaniewski, *Modern Pavement Management,* Krieger Publishing Company, Malabar, Fla., 1994.

[Steinthal 84]   B. G. Steinthal, "Infrastructure Maintenance Strategies— Governmental Choices and Decision Methodologies," *Infrastructure—Maintenance and Repair of Public Works, Annals of the New York Academy of Sciences,* Vol. 431, Dec. 1984, pp. 139–154.

[Urban 84c]   H. P. Hatry and B. G. Steinthal, *Guides to Managing Urban Capital Series,* vol. 4: *Guide to Selecting Maintenance Strategies for Capital Facilities,* The Urban Institute, Washington, D.C., 1984.

IMS Development and Implementation, Examples, Future Directions

Concept of Integrated Infrastructure Management Systems (IIMS)

16.1 Background

It is possible to develop a management system for a particular type of facility, such as pavements or bridges. Thus, it is reasonable to assume that an integrated management system can be developed to cover two or more similar types of facilities. The purpose of this chapter is to illustrate the validity of this assumption, and to show how it can lead to success.

The objective of implementing an integrated IMS within an agency, public or private, is to provide decision-makers with processed quantitative data they can use to examine the impact of various alternative scenarios. It represents an organized approach to helping them manage the infrastructure more effectively and efficiently.

Various elements of management systems are already in place in many agencies, but they are often not coordinated in a comprehensive and systematic fashion. A coordinated set of management subsystems can be used to establish formal procedures for recommending candidate projects and evaluating different strategies for solving problems, correcting deficiencies, and assessing trends of future needs. To tie all of these factors together, an Integrated Infrastructure Management System (IIMS) must incorporate forecasting models to develop trends of conditions, assess needs, and analyze future funding or budget scenarios. An IIMS can help the decision-maker in the development of both short- and long-term solutions.

With a history of several decades, the Pavement Management

System (PMS) has provided important and valuable guidance for IMS development in three general areas. These include:

1. State transportation systems (as prescribed in U.S. federal guidelines)
   - Pavements
   - Bridges
   - Safety
   - Congestion
   - Public transportation
   - Intermodal facilities

2. City infrastructure; same as for state transportation systems, with the addition of
   - Water
   - Sewer
   - Traffic signals, signs, and markings
   - Emergency services
   - Electricity
   - Garbage collection
   - Recycling
   - Drainage
   - Park facilities

3. Major unitized facilities—public and private
   - Airports
   - Nuclear power plants
   - Refineries
   - Parks and recreation areas
   - Other

After describing the framework for appropriate integrated systems, these three general application areas are discussed in the following sections.

16.2   Framework of Integrated Systems

As the various decision-support systems are developed and mature with use, there is a need to consider ways in which they can and should be coordinated. As well, there is a need to determine how the results should be integrated into information for strategic planning as a part of the executive management process using an Executive Information System (EIS). In general, individual systems (PMS,

BMS, traffic monitoring, and so forth) have been developed with only informal coordination among the various system developers. Correcting such a major deficiency by using integrated systems will provide policymakers with a tool for investing transportation and related infrastructure dollars in the most economically efficient manner. Transportation policy, for example, is no longer developed in a vacuum in most countries, including the U.S. Other state and national priorities, such as economic development, are directly influenced by transportation decisions and vice versa.

Multiple-criteria decision problems are best analyzed in an interactive manner, allowing the judgment of the decision-makers to be used directly. Such interaction permits decisions based on an assessment of relative allowable trade-offs among criteria in the region of feasibility. Multivariate analysis techniques have been developed to assist in this type of analysis for industry [Hoskins 88, Hsiung 91]. Sufficient structure and details need to be incorporated into the IMS so that the system represents the physical environment reasonably well, but detail should be limited to reduce the burden placed on the decision-makers.

A general framework for infrastructure management was presented in Chapter 1 (see Figure 1.9). An IMS using minimal assumptions will permit decision-making via any desired technique using the response information from the total management system, or on any portions thereof. Screening or ranking of acceptably good alternatives can be based either on qualitative or quantitative responses, depending on the type of information available, the degree of sophistication of the decision-makers, and the degree of detail considered essential in the management system.

16.3 Common Aspects of Management Systems

Good management systems have a number of things in common. These include analysis modules, a central database that is easily updated, compatible economic-analysis models, optimization models, and possibly a graphical interface with a GIS (geographical information system). Also, any infrastructure or facilities management system must allocate resources between capital improvements and maintenance, as discussed in Chapter 15. These common aspects of management systems are briefly discussed as follows.

16.3.1 Central database

A good database and data-processing system is the heart of any management system. A database is not a management system itself, but is required for one. A management system can only operate on the data made available to it. Systems that are set up independently with separate databases often duplicate many data items, but, more importantly, they are likely to have inconsistent location and identification information. An integrated management system operating on a coordinated central database, which contains all data needed for any of the management systems or subsystems, eliminates duplication and inconsistencies. The database must be flexible so that when the modular analysis components are upgraded and/or new components are added, it can also be updated or modified to accommodate the data needed for the new subsystems or the changes. The database should also be designed to accommodate expected additions of new infrastructure types. See Chapters 4 and 5 for details on database design and development.

16.3.2   Modular analysis tools

All management systems must have analysis tools that can manipulate the pertinent data from the central database to produce useful information and recommendations for management. Such analysis tools include the generation of summary statistics and graphs, economic analyses, benefit-cost analyses, decision methodologies, optimization routines, statistical analyses, and deterioration-rate analyses. An integrated system should have the flexibility to use modular analysis tools that can be easily updated as new technologies are developed. These tools should be consistent across the various infrastructure components that may be managed under the integrated IMS. Such consistency would include common or compatible optimization methodologies and life-cycle cost analyses.

Modularity of a system requires that the analysis tools be independent of the database. Unit cost information, deterioration rates, infrastructure types, functional classes, and so forth are all kept updated in the central database. The analysis tools are defined relative to the data needs, the models or analyses are performed, and the outputs are returned to the management system. As new modules are developed, each of these three main aspects must be included considering the rest of the system. The inputs are extracted from the central database, the analyses are performed in modular subroutines, and the control, interaction, and output review is performed at the user-interface level. Consistency in data needs and interface and output types is important to maintaining a seamless, integrated system.

16.3.3   Geographically based location/identification

Geographical Information Systems (GIS) have come of age, and are seeing widespread adoption around the world. Today, there is little reason that any integrated management system should be developed without using a GIS. A GIS provides a coherent, common geographical database location-referencing system, and a platform for relational databases that will minimize errors and confusion in collecting, processing, and storing data.

16.3.4   User-friendly graphical interface

The outputs from any management system are critical to its successful use. Long, detailed printouts or tabular data are not easily used by managers in making their decisions. Current systems have capabilities that include on-screen push-button controls, graphical views of the work environment, visual outputs and reports, on-screen photographs and videos, sound, and other utility improvement features. Such graphical interfaces not only enhance the use of the IMS output, they actually make the system more powerful and enjoyable to use.

16.3.5   Compatible benefit-cost economic analyses

Any infrastructure management system requires a valid life-cycle cost and/or benefit analysis. There is no reason that such analyses should not be consistent across all infrastructure subsystems within an agency, such as state, city, private entity, or airport. Using an IMS will help ensure economic consistency, including unit cost, definition of benefits and effectiveness criteria, discount rates, analysis period, definition of user and agency costs, level of service criteria, and goals.

16.3.6   Global consideration and allocation of resources

If an agency has separate individual management systems, then they allocate resources within the agency separately for individual subsystems. Thus, separate budgets are set up for water, sewer, pavements, traffic control, bridges, and so forth. This will likely result in suboptimization of the total budget, even though expenditures within a specific facility category may be optimal. An IMS should provide global consideration and allocation of resources. Certain constraints exist, however, for funding "set-asides," such as bridge, safety, and other improvement programs. These can be handled as constraints within the individual subsystems.

16.3.7   Evaluation of maintenance versus capital improvements

All facilities and infrastructure eventually deteriorate and ultimately become functionally or structurally unserviceable. Structural deterioration can produce major safety and operational problems. Facilities that are in poor physical condition present aesthetic, operational, and safety concerns. Management systems must be able to recommend budget division into timely maintenance or needed capital improvements. This requires accurate modeling and analysis of the costs and effects of maintenance activities that extend life or preserve facilities so they can be compared against capital-improvement replacement costs.

16.4   Common Aspects of PMS and BMS

The team that originally developed much of the pavement management philosophy, particularly the original, innovative thinking, was also selected by the National Cooperative Highway Research Program (NCHRP) in the U.S. to undertake the first definition of a Bridge Management System (BMS) [Hudson 87, Hudson 81, NCHRP 73]. The same concepts and application of system technology apply, even though the specific details vary. Both systems involve a change in the way an agency does business. Each requires a clear understanding of inputs, application models, and outputs of the system. Both systems require the use of clear models to estimate the behavior of the facility and models to evaluate cost. Each requires a system output function that can be optimized in relation to costs and benefits.

In essence, the development of BMS made extensive and efficient use of the earlier PMS development and experience. This lesson should be applicable to other individual management systems and to integrated infrastructure management systems.

16.5   State-Level Management Systems

State-Level Infrastructure Management Systems (IMS) are generally at a fairly high level of development. As described previously, pavement management system development and implementation experience has provided valuable guidance in terms of both concepts and application development. An integrated system framework should provide the umbrella under which thoroughness and consistency can be assured. Each of the six management systems prescribed by U.S. federal guidelines, as listed in Sec. 16.1, can be a subsystem in an overall IMS framework. They should be able to operate from a central database that contains all data for each individual management system.

Sinha and Fwa [Sinha 87] have outlined a concept of a total highway management system in which they envision a comprehensive highway management system as a three-dimensional matrix, as depicted in Figure 16.1.

The three dimensions are highway facility, operational function, and system objective. Table 16.1 lists the possible elements of each of

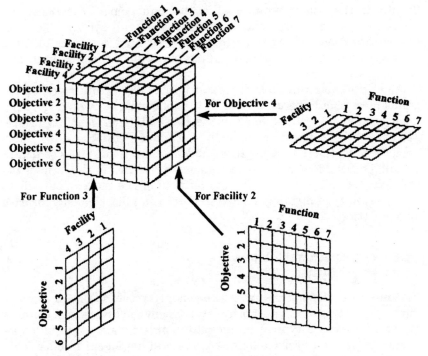

Figure 16.1 Three-dimensional matrix structure of a Highway Management System. [Sinha 87].

TABLE 16.1 Elements of Highway System Dimensions [after Sinha 87]

Elements of the highway facility dimension	Elements of the operational function dimension	Elements of the system objectives dimension
Pavement	Planning	Services
Bridge	Design	Condition
Roadside	Construction	Safety
Traffic-control devices	Condition evaluation	Cost
—	Maintenance	Socioecononomic factors
—	Improvements	Energy
—	—	Data management

the three dimensions. The suggested three-dimensional matrix structure indicates that a highway agency has a number of facilities to be managed in the highway system. The overall objectives of the agency could be well defined, and the effort for accomplishing these objectives divides the management task into a group of functions. Each facility in the system requires all of the management functions, and through these management functions the overall system objectives are accomplished. Using the framework presented in Figure 16.1, the highway management process can certainly be viewed as multiple-objective in nature.

An initial integrated IMS for states in the U.S., according to federal guidelines, should be comprised of the following individual management systems (or subsystems in the IMS): (1) pavement; (2) bridge; (3) safety; (4) congestion; (5) public transportation; and (6) intermodal—plus others that might be added subsequently. The structure should be flexible enough to accommodate changes in the form of the subsystems and in the inputs and outputs of the models, as well as the possibility of dividing a subsystem into smaller elements for ease of simulation.

16.6   Municipal Infrastructure Management Systems

An integrated infrastructure management system is equally important at the city or local authority level. Given the diversity of services provided within a typical municipality, an infrastructure management system can assist administrators and engineers to effectively manage and maintain services. This section illustrates frameworks that have been used for municipal infrastructure management systems (MIMS) with primary municipal subsystems. The discussion is presented on a conceptual level to provide a widely applicable outline.

One way to visualize the integration and linkages among the different subsystems, according to the Texas Department of Transportation (TxDOT), is shown in Figure 16.2. A representation of model interactions for the various subsystems is shown in Table 5.1. There is a strong link between public transportation and bridge and pavement management. A good illustration is provided by larger cities, where many arterial streets are being damaged by heavily loaded buses. In the case of Metropolitan Planning Organization (MPO) interactions, there will be particularly strong links among bridges, pavement, congestion, safety, and public transportation.

16.6.1   Systems definition for MIMS

One definition of municipal infrastructure, in terms of public works, is given by the American Public Works Association (APWA) [Stone 1974] as follows:

> Public works are the physical structures and facilities that are developed or acquired by public agencies to house governmental functions and pro-

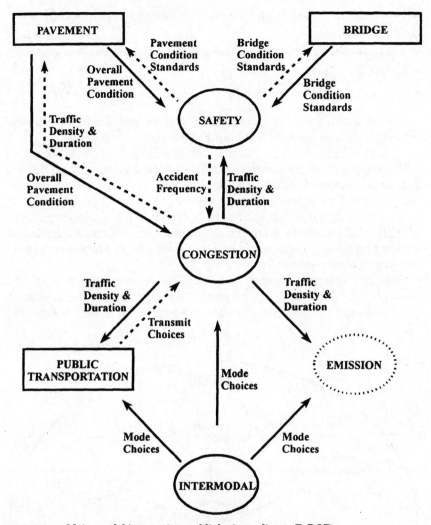

Figure 16.2 Major model integration and links (according to TxDOT).

vide water, power, waste disposal, transportation, and similar services to facilitate the achievement of common social and economic objectives.

The APWA goes on to list 18 categories of public works and environmental facilities that are included in this definition. Six categories of municipal infrastructure that are clustered together by industry and professional interest group are listed as follows:

- Roads group (roads, streets, and bridges)
- Transportation services group (transit, rail, ports, and airports)

- Water group (water, waste water, all water systems including water ways)
- Waste-management group (solid-waste management systems)
- Buildings and outdoor sports group
- Energy production and distribution group (electric and gas)

The potential comprehensive MIMS system and the identified subsystems linkages are illustrated in Figure 16.3.

Almost all cities deal with this set of subsystems, most or all of which normally are owned or controlled by the municipality. They are commonly large enough to require network-level oversight. In this manner the five most important systems, as shown in Figure 16.3, should be selected for initial development. With a comprehensive MIMS, all subsystems would be developed and incorporated into the system; however, it is possible to start with two or three and extend the coverage incrementally.

Once the system elements and boundaries have been defined, the next step is definition of the goals and objectives. Goals can be defined as the idealized "end states" toward which a plan is expected

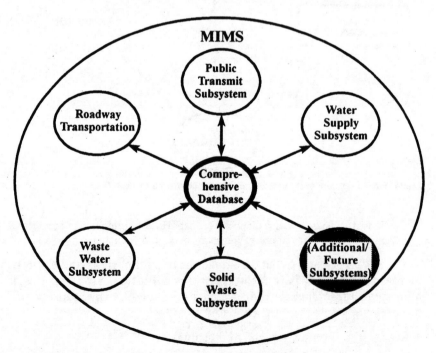

Figure 16.3 MIMS system and subsystems identified by APWA.

to move. Objectives are operational statements of goals. They must be measurable and attainable, where attainability is defined without reference to availability of resources or to budgetary restrictions. By definition, goals are broad and general in nature, while objectives are precise statements of what is to be done to reach a goal. One perspective on the goals and objectives of a MIMS can be obtained through the definition of the goals and objectives of each of the subsystems identified, the overall goal being the achievement of "common social and economic objectives."

The next step in the systems-development process is to create subsystem models. This involves the formulation, calibration, and validation of a basic model for each system or subsystem. It should be designed to permit prediction of the consequences of alternative plans, without the necessity of direct experimentation on the real-world system. At this point, the basic requirements for predicting the consequences of various alternative plans should have been developed. However, before generating alternatives, a set of criteria is required to assess goal achievement of the alternative plans.

To facilitate understanding of the requirements, models, processes, and so forth involved in each subsystem, some kind of basic systems analysis procedure, or analysis flow chart, should be developed for each of the subsystems shown in Figure 16.3.

Efficient collection and dissemination of useful and timely information is of critical importance to managing any enterprise. The most important issue for developing an overall infrastructure management system is the creation of a centralized database that allows a flow of information to and from each subsystem, and also from each activity as needed. This results in a more comprehensive, efficient, and timely flow of information for the MIMS as a whole, and for each subsystem in particular.

Although the specific information requirements of each subsystem will vary, certain broad information concepts apply to all subsystems. Each of the subsystems requires section identification information, preferably a common system that allows cross-referencing among subsystems. Each also requires historical data on system performance and system demand. Each requires information on subsystem-specific particulars, such as design and construction parameters, geometric data, and maintenance records. Finally, each subsystem requires tracking of economic effectiveness and both agency and user costs of service.

Ease of integration into one common municipal database is a consideration in the design of these data-requirement elements. For future development, it is suggested that data should also be classified as "required" or "desired."

16.6.2 Geographical information system (GIS) in MIMS

In recent years, Geographic Information Systems (GIS) have been widely used in various management fields, such as natural-resource management, wildlife protection, and environment protection. GIS has also been applied to several areas of infrastructure, including transportation systems and components, such as pavements and bridges. GIS applications for the Urban Roadway Management System (GIS-URMS) have been investigated at the University of Texas [Zhang 93, Zhang 96]. This work demonstrates the strong potential of GIS as a platform for developing an integrated overall infrastructure management system.

A Geographic Information System (GIS) is a computerized data-base-management system designed to manage spatially defined data, such as shown in Figure 5.2. The database contains information on spatially distributed entities that occur as points, lines, or polygons; and tools for capturing, storing, retrieving, displaying, interrelating, and analyzing locational as well as nonlocational data. Details are provided in Chapter 4.

There are three elements in a GIS: the geographic database, the attribute database, and the georelational data structure.

The *geographic database* processes three inherent characteristics: position, description, and variation with time. It can be structured in either grid (raster), vector, or triangulated irregular network (TIN) format.

The *attribute database* deals with all descriptive nongeographic information (variables, names, and characteristics) that identify a geographic feature. For instance, a lake can be identified by such attributes as its name, water level, water quality, chemical composition, salinity, and water temperature.

The *georelational data structure* deals with one of the main function of a GIS, which is to establish the relationship between the location of features in the geographic database and their corresponding descriptions in the attribute database. GIS performs the linkage between locational and attribute data by means of a georelational data structure.

16.6.3 MIMS central database

One of the important issues in developing an overall municipal infrastructure management system is data exchange and communication both among the subsystems and with other external systems. There are three well-known approaches to data management:

■ Relational approach

- Hierarchical approach
- Network approach

The relational database management systems (DBMs) available for microcomputers at the time of writing this book included dBase, FoxPro, Oracle, and Rbase. A relational DBMS can effectively perform data exchanges and communications among the proposed subsystems via common data fields, such as location ID, segment ID, or other items in the database. Figure 4.3 illustrates a possible integrated database for MIMS.

16.7 Unitized Facilities Management Systems

The third important area of application of overall infrastructure management is for what can be called "unitized facilities." These include, for example, airports, nuclear power plants, and major refineries. Each fulfills a major function, but requires a number of individual, independent subsystems to be operational.

For example, an airport requires roads, parking facilities, terminal facilities, aircraft parking aprons, taxiways, runways, and aircraft-control facilities. Each type of other such facility has a number of subsystems that must be combined to produce the required results. In this regard, the main function of an IMS is to integrate the various subsystems and their relationships to each other. The capacity of an airport, for example, can be controlled by any one of a number of subsystems. If access roads are inadequate and/or if parking is inadequate, then the overall airport capacity is reduced. The same can be said for terminal buildings, check-in facilities, baggage-handling facilities, number of aircraft gates, aircraft parking facilities, take-off and landing facilities, and air-traffic-control equipment. Poor management in any of these subsystems can limit airport capacity.

The concepts of infrastructure management can be applied to integrate airport subsystems effectively. This includes consideration of passengers, so that common database items include scheduling of take-offs and landings, related arrival of passengers, traffic delay, weather delays, weather information, passenger demand, and other factors.

16.7.1 Differences and common factors with other infrastructure facilities

Unitized facilities are in some ways easier—and in some ways harder—to handle than other management-system applications. Ease in handling arises from the fact that most facilities are located at a common location, such as an airport site. This is compounded when an agency, such as the Port Authority of New York and New Jersey, has to operate three airports within one municipal area. Nevertheless, the process of handling cohesive facilities in close proximity to each other offers a good application for infrastructure management.

Difficult aspects of the problem, which vary from, say, pavement or bridge management, include the broad diversity of functions that must be dealt with. In the case of airports, this ranges all the way from ground access to the airport through baggage handling to the complexity of air-traffic control. The scope of this chapter does not allow for detailed examination of these common and diverse factors. However, the benefits of unitized infrastructure management systems (UIMS) can be fully realized with a reasonable amount of research and development support for any of these areas.

Major items related to this concept include the potential use of geographical information systems within the facility itself to address the flow of passengers and the proximity of facilities. A relational database tying all of the factors together is also critically important to a unitized infrastructure management system.

A major factor in UIMS is the need for models that can describe the operation of each of the individual subsystems. However, this does not change the fact that such models are already required for operating the facilities without a UIMS. In other words, attempts to create a UIMS do not increase the difficulty in developing individual models. It does make explicit the need for such models, and it does require the management team to carefully consider the input and output of the individual models and their interactions. However, as previously discovered in bridge management and pavement management, while this explicit requirement may tend to make the current operating team uncomfortable, and seemingly apply a certain amount of pressure, it is very beneficial because it requires an explicit consideration of major factors.

In summary, UIMS is currently the least developed of infrastructure management systems. However, it offers the most potential. Private companies, such as Wal-Mart, have developed, to some degree, a UIMS for handling their inventory of goods and services. They also have used study methods to develop a common store configuration and common training methods so that uniformity and consistency will exist throughout their company over large geographic areas. It is recommended that consideration be given to UIMS development, particularly for public facilities like airports and parks, and facilities that require strong regulations, such as nuclear power plants. A UIMS can, for example, evaluate the overall aspects of the disposal of waste better than studying the parts independently. The steps involved include the following: (1) apply principles for general management systems as presented herein; (2) define all aspects of the facility; (3) properly model the system being managed; (4) develop a specific database for the system; (5) evaluate the problems that will occur; (6) build in monitoring procedures to evaluate problems; and (7) devise methods to correct them in the most cost-effective manner in terms of action and timing.

16.8   Implementation and Institutional Issues

Considerable effort has been invested in the development and implementation of pavement and bridge management systems. Nevertheless, as pointed out by many authors [Smith 91, Haas 94], the greatest problems associated with the successful implementation of pavement management systems are institutional issues. There are agencies that started development of a PMS in the mid 1970s that are still ineffective in its use. It is necessary to address the problem of institutional issues and the reluctance of agencies to change, to help ensure that the integrated systems developed in the future actually get implemented.

16.9   Example of an IIMS

One of the most ambitious and comprehensive IIMS initiatives at the time of writing this book has occurred in Kuwait [Abdel Kader 96, Al-Kulaib 97]. The main individual systems in the IIMS, as developed for The Ministry of Public Works and The Ministry of Water and Electricity, include the following (see also Figure 16.4):

I. Roadway Systems
   1. Pavement management system (PMS)
   2. Bridge management system (BMS)
   3. Right-of-way features management system (RFMS)

II. Utility Systems
   1. Sanitary-sewer network management system (SANS)
   2. Storm-sewer network management system (STMS)
   3. Watermain network management system (WNS)
   4. Electrical/communication management system (ECMS)

III. System Integration
   1. Spatial data management (SDM)
   2. Work management system (WMS)

Key features of Kuwait's IIMS include the following [Abdel Kader 96]:

- Total interpretation of the various system modules at both the data and user-interface levels

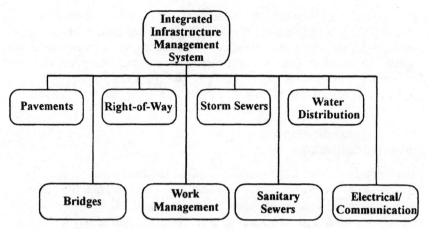

Figure 16.4 Scope of Kuwait's IIMS.

- Use of fourth-generation language (4GL) database managers that are standard query language (SQL) compliant
- Implementation of the IIMS in a client/server environment
- Database independence of the system modules, so that client applications can access data in various 4GL databases
- Ability of client applications to utilize a graphical user interface (GUI) in any of Microsoft windows, MOTIF/x-windows, or Macintosh
- A comprehensive IIMS database containing data fields for all infrastructure components
- Capability of managing data on an integrated global basis
- Capability of producing a comprehensive set of understandable tabular and graphical reports, including color-coded maps and digitized images
- System security, to restrict user access at several levels
- Compatibility with existing GIS, Management Information System, and Maintenance Management System
- Capability of network analysis on any individual component system and on a multiple-system basis
- Ability to operate on both a local-area network and a wide-area network
- Open architecture to allow future expansion
- Capability of operating with minimum response times
- Capability of assessing the condition/performance of each system component individually and collectively

- Ability to produce optimal multi-year priority programs of rehabilitation and maintenance for each IIMS component system and for the IIMS as a whole
- Through the work-management system, the ability to process service requests, work-load analysis, performance tracking, and work-order historical information
- Ability to automatically adjust maintenance and rehabilitation strategies to varying service levels
- Ability to automatically identify and assess combined projects (e.g., roads and sanitary)
- Interactive with the budget process
- Ability to manage historical information
- Ability to provide a fully integrated, network-level electronic map of any individual infrastructure to be interfaced with any future GIS vendors
- Capability of receiving input data and producing reports in both Arabic and English

References

[Abdel Kader 96]   M. O. Abdel Kader and A. A. Al-Kulaib, "Kuwait Integrated Infrastructure Management System," *Proceedings, International Road Federation Asia-Pacific Regional Conference,* Taipei, Nov. 1996.

[Al-Kulaib 97]   A. A. Al-Kulaib, M. O. Abdel Kader, M. A. Karan, J. Powell, and G. Johnston, "Integrating Road Management with Other Infrastructure Assets Management in Kuwait," prepared for the International Road Federation XIII World Meeting, Toronto, Ont., June, 1997.

[Haas 94]   R. Haas, W. R. Hudson, and J. P. Zaniewski, *Modern Pavement Management,* Kreiger Publishing Company, Malabar, Fla., 1994.

[Hoskins 88]   J. C. Hoskins and D. M. Himmelblau, "Automatic Process Control Using Reinforcement Learning in Artificial Neural Networks," *Proceedings, 1st Annual Meeting, International Neural Network Society,* 1988.

[Hsiung 91]   J. T. Hsiung and D. M. Himmelblau, "Development of Control Strategies via Artificial Neural Networks and Reinforcement Learning," *Proceedings 1991 ACC,* Boston, Mass., 1991.

[Hudson 81]   W. R. Hudson and R. Haas, "Development, Issues, and Process of Pavement Management," *Pavement Management: Proceedings of National Workshops, FHWA-TS-82-203,* Federal Highway Administration, Washington, D.C., June 1981, pp. 25–35.

[Hudson 87]   S. W. Hudson, R. F. Carmichael III, L. O. Moser, W. R. Hudson, and W. J. Wilkes, "Bridge Management Systems," *NCHRP Report 300,* Transportation Research Board, Washington, D.C., 1987.

[NCHRP 73]   W. R. Hudson and B. F. McCullough, "Flexible Pavement Design and Management—Systems Formulation," *NCHRP Report 139,* Highway Research Board, National Research Council, Washington, D.C., 1973.

[Sinha 87]   K. C. Sinha and T. F. Fwa, "On the Concept of Total Highway Management," in *Transportation Research Record 1229,* Transportation Research Board, National Research Council, Washington, D.C., 1987.

[Smith 91]   R. E. Smith, "Addressing Institutional Barriers to Implementing a PMS," *Pavement Management Implementation,* American Society for Testing and Materials, Philadelphia, Pa., 1991.

[Stone 1974]   D. C. Stone, *Professional Education in Public Works / Environmental Engineering and Administration,* American Public Works Association, Chicago, Ill., 1974.

[Zhang 93]   Z. Zhang, W. R. Hudson, T. Dossey, and J. Weissman, *GIS Applications in Urban Roadway Management Systems,* Center for Transportation Research, The University of Texas, Austin, Tex., April 1993.

[Zhang 96]   Z. Zhang, "A GIS Based and Multimedia Integrated Infrastructure Management System," Ph.D. thesis, The University of Texas, Austin, Tex., Aug. 1996.

17

Visual IMS:
A Working Infrastructure
Management System
and Applications

17.1 Introduction

In previous chapters, various concepts and components of infrastructure management systems were discussed. The most important step required to gain the real benefits of IMS, however, is to apply these concepts, and integrate the components into an active working system. Although no comprehensive working IMS will be perfect at its initial stage of development, it is the essential starting point from which subsequent improvements can be made.

This chapter introduces Visual IMS, a working GIS-based and multimedia-integrated infrastructure management system that was developed at the University of Texas at Austin [Zhang 96]. The general description, main structure, and major operational functions of Visual IMS are presented, along with some data-flow diagrams and computer-screen examples.

17.2 General Description of Visual IMS

Visual IMS is a GIS-based and multimedia-integrated infrastructure management system developed primarily for application by public-works agencies. Currently, three subsystems are included: roads, water supply, and waste water. The main purpose of Visual IMS is to provide a series of rational, well-ordered analyses of input data so that all infrastructure involved can be effectively maintained above the specified minimum level of service at a minimum overall cost. Visual IMS is programmed in Avenue, an object-oriented GIS programming language capable of handling the operations of both numerical data and graphical objects. Visual IMS runs under the Windows environment, and is facilitated with user-friendly graphical user interfaces (GUIs). It can be operated as a stand-alone IMS, or used to enhance the user-friendliness and visual capability of any other IMS. The key difference between the Visual IMS and other decision-support systems is that Visual IMS is able to handle a wide variety of data and information visually, analyze them spatially, and present the results graphically, taking advantage of both GIS and multimedia technologies.

The operation of Visual IMS is divided into two parts: (1) analytical operations based on decision models; and (2) visualizing/mapping operations based on GIS and its integration with multimedia.

17.3   Analytical Operation of Visual IMS

The analytical functions of Visual IMS are divided into six categories, which are implemented as six function modules: (1) database; (2) model setup; (3) performance evaluation; (4) planning; (5) budget allocation; and (6) proximity analysis. Figure 17.1 shows the six categories and the specific functions under each category. Each subsystem (road, water, or waste water) has its own separate modules for the first four categories. The budget allocation and proximity analysis modules are for the entire system, and include all three subsystems in each, respectively.

17.3.1   Database and data input

A database is an essential component of any decision-support system, such as Visual IMS. Five groups of input data are required by the Visual IMS program: (1) geographic data that define the coordinate location of particular infrastructure features; (2) inventory data that provide an accounting of the existing physical infrastructure features; (3) condition data that are essential for conducting various performance-related evaluations and analyses; (4) policy and cost data (in terms of weighting factor and unit costs); and (5) multimedia data, such as videos and images, for visualizing infrastructure conditions.

There are primarily two ways to prepare the required database. One is to convert an existing database into a Visual IMS compatible database; the other is to create a new database and fill it with data collected in the field. Figure 17.2 is a typical data-input window in Visual

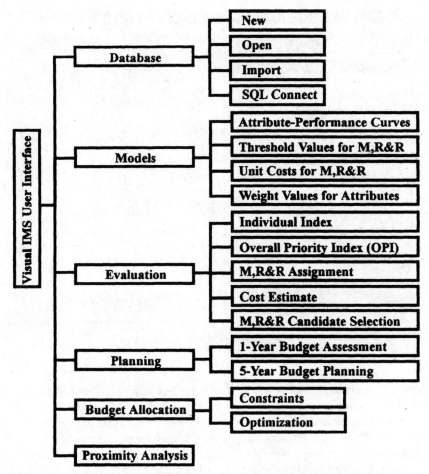

Figure 17.1   Six categories of analytical functions in Visual IMS. [after Zhang 96].

IMS. For economic reasons, users should always examine the existing database for use with Visual IMS before carrying out field data collection. With its open interface for data import and export, Visual IMS is capable of accommodating a wide variety of existing data. Data stored in any other SQL-compliant database can also be accessed by Visual IMS through its SQL connection, as illustrated in Figure 17.3.

17.3.2   Model setup

As explained earlier, Visual IMS was designed and developed as a generic system rather than an application-specific system. The models implemented in Visual IMS are basically generic models. Users

Figure 17.2 A typical data-input screen [after Zhang 96].

should exercise caution when these models are employed with their default parameters. The purpose of model setup is to calibrate and/or replace the default parameters and curves using local experience and practice. Curves and parameters that usually require modification or update during model setup include: (1) membership curves between infrastructure attribute value and normalized performance value; (2) performance threshold values for M,R&R; (3) unit costs for M,R&R; and (4) weight values for infrastructure attributes. To ease the process of model setup, Visual IMS is facilitated with user-friendly graphical model editors, as illustrated in Figure 17.4. These graphical model editors allow users to interactively modify model curves on the computer screen.

17.3.3 Performance evaluation

One of the main uses of performance evaluation is selecting and ranking candidate projects for M,R&R [NCHRP 79, Haas 94]. The inputs for the evaluation module include: (1) attribute measurements of infrastructure [such as Pavement Condition Index (PCI) for the road subsystem]; (2) unit costs for M,R&R alternatives; and (3) available

Figure 17.3 An example of establishing SQL connection to retrieve data. [after Zhang 96].

budget. Figure 17.5 illustrates the process of selecting M,R&R candidates according to the Overall Priority Index (OPI) model. An OPI model is developed by considering a weighted combination of all of the attributes that are important to the M,R&R decision-making process. The M,R&R candidate selection starts with an OPI evaluation, which uses the OPI model and the condition data stored in the attribute database. A maintenance-and-rehabilitation action class is then assigned to each M,R&R section based on its infrastructure condition index value. The cost of the assigned M,R&R is calculated for each section.

If the total estimated cost for all such sections is within the available budget limit, M,R&R can be done on all sections as they are needed. Otherwise, the M,R&R sections are prioritized by their OPI values. A final list of M,R&R candidate projects is generated by

Figure 17.4   Graphical model editor in Visual IMS. [after Zhang 96].

adding the section with the next higher priority to the list until the budget limit is reached.

17.3.4   Budget planning

The budget-planning module is based on the infrastructure performance index (PI) and the performance-prediction model presented in detail by Zhang [Zhang 96]. Figure 17.6 shows the analysis process for budget planning. The algorithm starts by calculating the length-weighted average of performance for each infrastructure system. The calculated average PI is then compared with the user-desired PI. If the calculated PI is below the desired PI, the algorithm proceeds to assign M,R&R actions, calculate costs, prioritize project sections, and re-evaluate the PI values. The total cost is then written to the output report. If the calculated PI is higher than the desired PI, the algorithm proceeds directly to predict the PI for the next year. This process is repeated until it reaches the last year of the five-year planning period. As an example, the typical input and output values for a five-year budget-planning analysis are shown in Figure 17.7.

Figure 17.5   The process of M,R&R candidates selection. [after Zhang 96].

The inputs are the average performance levels expected by users for the planning period; the outputs include the M,R&R candidate projects and the budgets needed for achieving the specified performance. The output of yearly budgets is presented in both tabular and graphical formats.

17.3.5 Budget allocation

The objective of the budget-allocation module is to maximize the overall performance of the infrastructure system with the constraint that the available budget is limited [Aguilar 73]. Figure 17.8 shows the analysis process for budget allocation. The budget-allocation analysis is based on a set of curves stored in the model base. Each curve defines a relationship between the available budget level and the corresponding level of performance for a particular infrastructure subsystem. The relationship curves are constructed with the infrastructure condition data in the attribute database, and need to be

Figure 17.6 Analysis process of budget planning. [after Zhang 96].

recalibrated whenever a change is made to the condition data. To ensure that a particular subsystem receives at least a minimum amount of money, users can specify a minimum budget level for each subsystem. Of course, the sum of the minimum budgets must not exceed the total available budget. The optimization algorithm then performs the analysis and reports the results. Figure 17.9 shows the typical input and output of an optimizing budget-allocation analysis among different infrastructure subsystems.

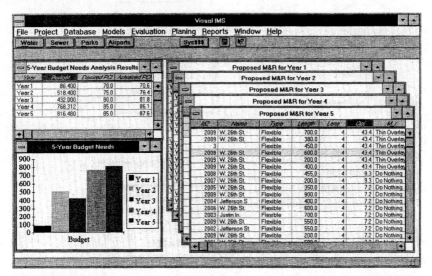

Figure 17.7 Typical input and output of a five-year budget-planning analysis. [after Zhang 96].

17.3.6 Proximity analysis

One of the most important spatial-analysis functions in Visual IMS, proximity analysis, is aimed at improving the overall efficiency of M,R&R. The most effective approach to achieve this purpose is to schedule all M,R&R activities for the related infrastructure systems in a systematically coordinated manner. In this way, all M,R&R works that share the same geographical location can be planned and carried out together in the most efficient and economical sequence. For example, any problem associated with the waste-water pipelines should be identified and fixed before applying an overlay in that GIS section. The inputs for proximity analysis include: (1) geographical data that defines the geometric location of the infrastructure involved; and (2) M,R&R candidate projects obtained from the evaluation module. The output from proximity analysis is the M,R&R status of those infrastructure features that are connected with a specific infrastructure feature scheduled for M,R&R.

Figure 17.8 Analysis process of the optimization of budget allocation. [after Zhang 96].

17.4 Visualizing/Mapping Operation of Visual IMS

Another important part of the Visual IMS program is the visualizing/mapping operation. It enhances the process of analytical operation and supports the presentation of analysis results. As shown in Figure 17.10, visualizing/mapping functions that are useful for infrastructure management can be classified into six categories: (1) map display; (2) feature selection; (3) spatial query; (4) spatial analysis; (5) multimedia link; and (6) map production.

Figure 17.9   Typical input and output of optimizing budget allocation. [after Zhang 96].

Map display.   Map display includes a wide variety of operations for map manipulation, such as zooming into a specific area of the map for a closer look at certain infrastructure features and editing the map legends to produce the best visual effect. New infrastructure systems can be added to the system through the map-adding/deleting function. Furthermore, infrastructure attribute data can be classified and color-coded on the map for better presentation, as illustrated in Figure 17.11.

Feature selection.   Feature selection allows some of the infrastructure features to be selected interactively for analysis. The selection can be done with the mouse, graphic shapes drawn on the theme, records in a table associated with the map, or even a query operation.

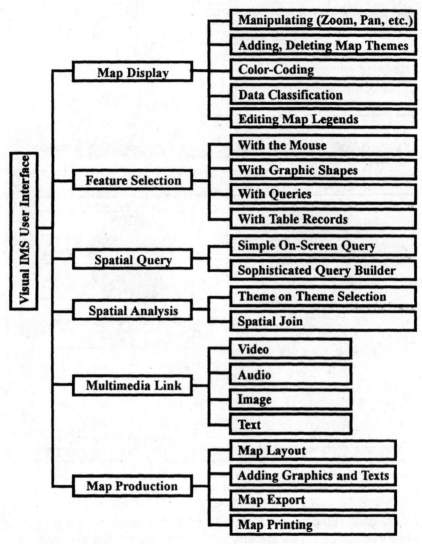

Figure 17.10 Six categories of visualizing/mapping functions in Visual IMS. [after Zhang 96].

Spatial query. Spatial query is useful in examining information related to a specific infrastructure feature or a set of features that meet certain criteria. For example, the sophisticated query builder can help to find all the pavement sections with a PCI less than 55 and a traffic volume (AADT) greater than 35,000 and highlight them.

Spatial analysis. "Theme-on-theme selection" and "spatial join" are two of the spatial-analysis functions available in Visual IMS. The

Figure 17.11   Highway base map and related pavement information. [after Zhang 96].

proximity-analysis module discussed earlier is built on the basis of theme-on-theme selection.

Multimedia link.   Multimedia link allows multimedia information, such as video, audio, image, and text, to be attached to the corresponding infrastructure map features. The attached multimedia information can then be accessed and used simultaneously through an on-screen query, as illustrated in Figure 17.12.

Map production.   The effective presentation of analysis results is an important part of IMS. With Visual IMS, analysis results and related information can be produced as tables, charts, and maps, or as a combination of forms with a customized layout. The final document can be exported as a graphic file and/or printed as a hard copy.

17.5   Hardware, Operating System, and Memory

For successful installation and operation of Visual IMS on a personal computer (PC), the computer must meet the required minimum configuration outlined below, operate under the required operating system,

Figure 17.12 Multimedia information capability in Visual IMS. [after Zhang 96].

and have adequate hard-disk space, random access memory (RAM), and virtual memory. These requirements are discussed as follows.

Hardware requirements. The hardware requirements include computer CPU (central processing unit), hard disk, monitor, pointing device, sound card, and speakers.

- *Computer CPU* The computer should be IBM-compatible and have at least an 80386 or higher microprocessor. It also should have a hard disk and a 3.5-in floppy-disk driver.

- *Hard-disk space* The installation of Visual IMS requires at least 20 MB of hard-disk space available on the computer.

- *Monitor* A video graphic adapter (VGA) or higher-resolution monitor is required.

- *Pointing device* A Microsoft mouse or compatible pointing device is required.

- *Multimedia device* Sound card, CD-ROM, and speakers must be properly installed and configured if the user wants to employ all the multimedia functions in Visual IMS.

Operating system requirement. To run Visual IMS on a PC, MS DOS 5.0 or later and one of the following Windows operation systems are required: Microsoft Windows 3.1, Windows for Workgroups 3.11, Windows NT 3.1, or their later versions.

Memory requirement. A minimum of 8 MB RAM is required to run Visual IMS. However, 16 MB RAM or higher is recommended for more efficient operation, especially if the user wants to run other applications simultaneously with Visual IMS. A minimum of 12 MB file-swap space also is required as permanent or temporary virtual memory.

Note that the requirements presented so far are for running Visual IMS on a stand-alone PC. To implement Visual IMS in a distributed network, such as a local-area network (LAN) environment, different hardware and software will be required.

17.6   User Training

The GIS-oriented nature and the multimedia-integrated features of Visual IMS are somewhat more complicated than simple PC programs, and it is expected that a medium level of user training will be required. Some basic knowledge of GIS will be helpful to personnel in understanding the concepts and principles of Visual IMS.

It is estimated that two days of training will be required for an engineer who has a normal level of computer knowledge (i.e., a person who understands a Windows operating system and knows how to run Windows-application software) to perform analyses with the basic functions in Visual IMS. More sophisticated applications with Visual IMS will require additional training, depending on the level of prior knowledge.

The primary purpose of this early version of Visual IMS is to demonstrate the concepts, approaches, and models developed under this study. All of the major function modules have been included in this version of the program, and when future work is invested, further improvements can be made to the Visual IMS program.

17.7   Application of Working Infrastructure Management Systems

Unless it is implemented or applied to solve practical engineering problems, an infrastructure management system has little value. This section presents several case studies involving the application of Visual IMS. These case studies demonstrate how an IMS can be used not only as a comprehensive management system, but also as an enhancement to existing management systems. The application of the IMS concept to other management problems is also illustrated with a case study.

Three case studies were developed as part of the effort to test the Visual IMS program and to manifest the concepts presented in previous chapters. Case Study I shows how an IMS can be implemented to work with an existing management system. Case Study II illustrates how the analytical capabilities of the IMS can be used to answer engineering questions in infrastructure management. Case Study III serves as an example of how the concepts of IMS can be used for other similar management applications.

17.7.1   Case study I: Montana PMS

An important implementation of IMS was carried out by the Texas Research and Development Foundation (TRDF) to support a pavement management system (PMS) it had developed for the state of Montana. As indicated earlier, the purpose of this case study was to demonstrate how an IMS could be used with an existing management system to improve the user-friendliness and visual capability of the system.

The TRDF PMS for the state of Montana is a Windows-based system that was developed with PowerBuilder, an SQL-compliant database management system [TRDI 96]. Although the PMS is capable of conducting various sophisticated analyses for pavement management, it does not have a GIS interface, which is increasingly demanded by users who want to present analysis results geographically. Major tasks for this case study included: (1) developing the interface between the IMS and the TRDF PMS; (2) preparing a base map of the Montana highway system; (3) relating pavement-attribute data records to their corresponding map features; and (4) incorporating multimedia data. Both desktop PC and laptop PC computers were used for this case study.

17.7.1.1   Interface between Visual IMS and TRDF PMS.   The main purpose of such an interface is to facilitate data communications between the two computer programs. Two methods were employed for this purpose. The first method was to export the analysis results from TRDF PMS as a dBase file to be read directly by the IMS. The results were used as input for the IMS. The second method was to establish an SQL connection between Visual IMS and the TRDF PMSs database so that the result tables of PMS analyses could be directly retrieved by the IMS.

17.7.1.2   Base-map preparation.   The "National Transportation Atlas Data Bases: 1995" (NTAD-95) was used to prepare the Montana base map. NTAD-95 was developed and distributed by the Bureau of

Original NTAD-95 data stored in fixed-digit format

Figure 17.13 Conversion of NATD-95 into a GIS-readable format.

Transportation Statistics of the U.S. DOT free of charge [USDOT 95]. The original data are stored as ASCII text files in fixed digit format— a format that cannot be directly used by any GIS software. To solve this problem, a computer program was written to subtract the longitude/latitude coordinates that define the Montana highway network from the ASCII file and to rewrite the subtracted coordinate data into another GIS-readable text file, as illustrated in Figure 17.13.

17.7.1.3 Relating pavement attribute to base-map features. Pavement sections are represented by the milepost method in the analysis results of the TRDF PMS for Montana. The relationship between the analysis results and their corresponding map features was established by creating a common ID for the pavement-data table based on its mileposts. The common ID was then used to either "join" or "link" the pavement data to their corresponding sections on the base map.

17.7.1.4 Multimedia data. Some multimedia data were also prepared for the case study, including: (1) traffic video in *.AVI format; (2) pavement surface image in *.BMP format; (3) audio comment in *.WAV format; and (4) information summary in *.TXT format.

17.7.2 Case study II: analytical capability

In Case Study I, the analytical part of the problem was largely handled by the TRDF PMS. The intention of Case Study II, therefore, is to demonstrate the analytical capabilities of Visual IMS with some common problems. Due to the limitation of time and resources, the data used for Case Study II approximates data, but is not real data. The infrastructure features on the base map were created through interactive on-screen digitizing, using an aerial photograph as a background. The database tables of infrastructure condition were developed by using the Visual IMS database module. All the six categories of analytical functions have been tested with the developed infrastructure condition database. An example of the typical input and output for a five-year budget-planning analysis are presented in Figure 17.7. The input for the analysis was the average performance levels expected by the user for the planning period; the output included the M,R&R candidate projects (highlighted records) and the budgets needed for achieving the specified performance. The output of yearly budgets was presented in both tabular and graphical formats. According to the outcome of this case study, all function modules have produced reasonable results.

17.7.3 Case study III: conceptual TMS for Cobb County DOT

Case Study III was carried out by the Texas Research and Development Foundation (TRDF) as a direct application of the concepts of Visual IMS. In December 1995, the Cobb County Department of Transportation (Cobb DOT) requested proposals for the development of a Transportation Management System (TMS) for their surface-transportation network. The Cobb DOT required six management modules to be included in the TMS:

1. Road inventory module
2. Traffic congestion module
3. Bridge maintenance module
4. Pavement management module
5. Traffic signal module

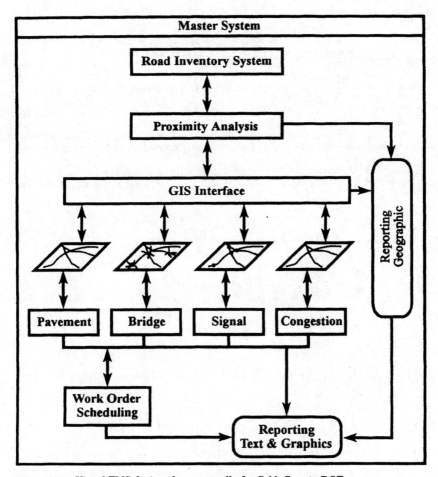

Figure 17.14 Visual TMS designed conceptually for Cobb County DOT.

6. Work order scheduling module

Since these six modules have to be integrated and operated under one master system, it was decided to use GIS as the integration platform. The objective of Case Study III was to conduct a conceptual design for the TMS. After analyzing the requirements of the Cobb County TMS, the concepts and frameworks of Visual IMS were found readily applicable to the development of a TMS. As a result, a Visual TMS was conceptually developed for the Cobb DOT. Figure 17.14 depicts the major function modules and their relationships in the Visual TMS designed for the Cobb DOT.

References

[Aguilar 73]   R. J. Aguilar, *Systems Analysis and Design in Engineering, Architecture, Construction, and Planning,* Prentice-Hall, Englewood Cliffs, N.J., 1973.

[Haas 94]   R. Haas, W. R. Hudson, and J. Zaniewski, *Modern Pavement Management,* Krieger Publishing Company, Malabar, Fla., 1994.

[NCHRP 79]   "Pavement Management System Development," *NCHRP Report 215,* Washington, D.C., Nov. 1979.

[TRDI 96]   *TRDI Visual PMS User Guide,* Texas Research and Development Inc., Austin, Tex., 1996.

[USDOT 95]   *National Transportation Atlas Data Bases: 1995, BTS-CD-06,* Bureau of Transportation Statistics, U.S. Department of Transportation, Washington, D.C., 1995.

[Zhang 96]   Z. Zhang, *A GIS Based and Multimedia Integrated Infrastructure Management System,* Ph.D. thesis, the University of Texas at Austin, Austin, Tex., Aug. 1996.

18

Benefits of Implementing an IMS

18.1 Introduction

Good IMS is not business as usual. Far too many people have jumped on the bandwagon and proclaimed to the world in various conferences and state meetings: "we are doing infrastructure management." IMS requires a change in the way some existing agency personnel do their thinking and day-to-day work, i.e., a change in paradigm. It is difficult to change; no one is initially comfortable with change. Information of all kinds is essentially raw data. It is abundant, sometimes cheap, easy to acquire, and sometimes hard to avoid or sift through. It is often provided in great quantities. Intelligence, by which we mean processed, evaluated, and meaningful data, is more difficult to acquire and more expensive, but is also much more important to IMS.

Models, methodologies, and procedures for IMS are aimed at providing the essential service of transforming raw data into information and intelligence, and essential service to administrators and decision-makers. It is aimed at reducing risk in policy and budgeting decisions with respect to particular projects. For such a process to be viable, several prerequisites must be fulfilled, namely:

- Decision-makers must appreciate the importance of this service: They should also be capable of understanding the usefulness and limitations of the process. There is a need in this area that offers a great opportunity to sell improved IMS.

- Infrastructure management systems must be available, usable, and credible. This is not always the case, as resources applied to the process have often been inadequate or mandated.

Public policies involving risks are most likely to be deemed acceptable when based on credible technical and engineering information, and where sound trade-off and impact analyses have been performed.

Most man-made systems (public and private), including infrastructure facilities, are planned, designed, constructed, operated, and modified or rehabilitated under conditions of uncertainty and risk. Thus, the assessment, quantification, evaluation, and analysis of risk and reliability in the context of systematic infrastructure management is essential to the process of good decision making.

18.2   Sources of the Skepticism on Infrastructure Management

While IMS has been around in one form or another for over two decades, there is still major skepticism at middle- and upper-management levels in public agencies. The following list summarizes some sources of this skepticism:

1. Insufficient planning and resources allocated for IMS development, maintenance, and upgrading of the process

2. Use of the model or the IMS is delegated to people who do not truly understand the process

3. Lack of incentives to properly implement and document IMS

4. Insufficient basic scientific research for the models

5. Misuse of models and incorrect application of systems

6. Overemphasis on use of computers as a "black box"

7. Proliferation of models and methods, and lack of systematic inventory of available methods

8. Improper calibration, validation, and verification of some IMS models

9. Lack of communication among developers, users, and beneficiaries of IMS

10. IMS must be recognized as "means" not "ends"

11. Lack of an interdisciplinary team in the development process. For example, lack of a statistician in many agencies

12. Strengths, weaknesses, and limiting assumptions of IMS often unrecognized by decision-makers

13. Insufficient data available to properly use the methods in early phases

14. Inadequate understanding of the true "costs" and "benefits" of IMS

How can one go about "selling" the infrastructure management system? The process of communicating ideas about IMS is as important as the content of the communication. Determining what change should occur (what to do) is easier than determining how to bring about that change (how to do it). Implementation of a change as great as the application of IMS requires a carefully coordinated and sequenced set of changes in training policy, personnel, and practice executed over a period of time. The process of implementing change is as important as the changes themselves, particularly if it is desired to get the changes adopted and implemented.

A number of organizational barriers exist that work against change in general, and against the adaptation of modern systems technology in particular. A social system, such as a public agency, tends to react to new changes in ways that keep old policies and practices stable. Major changes, such as developing and implementing an IMS, tend to provoke complex and compensating reactions. Unless the interrelationships between different divisions or parts of an organization are accurately identified, the organization may react to subvert the intended change. Specifically, to successfully implement an IMS one must

- Clarify issues of choice
- Fully inform the decision-makers
- Achieve substantial, effective, organizational agreement on a course of action that is feasible, equitable, and desirable

Considerable attention is needed as to the best way to communicate the benefits of IMS. There are several ways to do this, but it is clear that good, well-documented case-study examples of IMS are useful in communicating its benefits.

Any approach to the issues raised here requires the consideration of the costs as well as the benefits of performing infrastructure management and implementing an IMS. The issues of cost are treated in the following section. Benefits of various kinds are treated subsequently.

18.3   Costs Related to IMS

There are several types of costs associated with IMS. The two primary areas are these:

1. The costs of developing an IMS, along with the necessary and appropriate data for using the IMS and keeping it current—that is, the cost of the IMS process itself.
2. The true cost of the infrastructure itself; more correctly, the cost savings resulting from good management.

In reality, it is ultimate savings in actual life-cycle expenditures that reflect the effectiveness of IMS. The total investment and related costs must be considered, along with savings and benefits that are realized from the effective implementation of IMS.

Many problems occur in assessing either of the costs outlined above. Apparent costs can vary greatly, depending on accounting procedures and methodologies within a given organization. Some agencies do not consider overhead or indirect costs at all when they are carrying out activities within their agency. On the other hand, if the same activities are done by contract or the use of consultants, then indirect costs are clearly included in the bid or final contract price for the work. All cost figures must be considered on a common basis, and should include all costs both direct and indirect of doing the necessary work.

Obtaining cost information to be used in evaluating costs to benefits or cost-effectiveness is a difficult task. This is due partially to the fact that few agencies have a fully implemented IMS at the present time, and those that do often have not kept effective comparisons and records of development costs. In the case of the facilities themselves, the costs associated with construction can be documented, but they have usually occurred over many years, and a common basis of comparison is difficult to obtain. More substantially, the costs of maintaining infrastructure are extremely difficult to define, and few agencies have truly good maintenance cost information.

Several approaches can be considered. These involve pulling together costs on a semihypothetical basis for general comparisons. Case studies from individual agencies also can be useful. Particularly, cost information on pavement management in which contractor-related IMS activities have been carried out may be useful.

18.4 Benefits Associated with IMS

In Table 18.1, a variety of detailed products and benefits associated with IMS are presented. Some of these are couched in terms of elected representative benefits, senior management benefits (decision-makers), and technical-level benefits for operators of an IMS, including designers and maintenance personnel. Table 18.2 summarizes some of the benefits that occur at agency level (using highway examples). Table 18.3, on the other hand, lists many of the costs the agency will incur. Table 18.4 presents a case-study economic comparison, including costs and benefits of IMS. Although this is not totally comprehensive, it directly points out the factors involved, and shows the significant benefits that can build up [Haas 95].

A key issue with benefits is that some are quantifiable and measurable, but many are subjective and general in nature. Benefits must be

TABLE 18.1   Benefits and Costs that Occur with IMS

A. General Factors

Benefits involved	Costs involved
Realization of the magnitude of the investment that the agency has better chances of making correct decisions on spending funds Improved intra-agency coordination Improved use of technology Improved communication	Developing the system Data collection Actual operation of the system

B. Elected Representatives Receive

Benefits	Costs
Being able to defend/justify programs of maintenance and rehabilitation Having assurance that programs represent best expenditure of public funds Reduction of pressure (from constituents) to make arbitrary program modifications Getting objective answers to the implications of ■ Lower levels of funding ■ Lower standards	Development and implementation Assigning personnel Actual operation of the system

C. Senior Management Receives

Benefits	Costs
Comprehensive, comparative assessment of ■ Current status of the network ■ Expected future status Objectively based answers to ■ What level of funding is required to keep the current status? ■ The implications of greater or lesser budgets ■ The implications of deferred work ■ The implications of lower standards Being able to objectively justify capital spending and maintenance programs to the elected council, or legislature Having the assurance that the recommended program represents the best use of available dollars Being able to define the "management fee" as a percent of the spending on capital and maintenance work (i.e., the cost of the IMS divided by the spending)	Data collection Developing and installing the IMS Staffing On-going costs of operating the IMS Assigning personnel

brought to a common basis if they are to be compared with costs in a truly analytical way. A big problem with IMS today is that user benefits are almost totally excluded from the calculations of IMS decision-making. It was considered sufficient in the past to provide "adequate facilities" to serve the public. The idea of improved benefits accruing to the user based on better service and lower user operating costs has

TABLE 18.1   Benefits and Costs that Occur with IMS *(Continued)*

D. Technical Personnel Receive

Benefits	Costs
More explicit recognition of the various administrative and operating elements of the organization	Having to make changes in the ways things have been done
Increased awareness of available technology and how to use it	Possible extra time and effort to upgrade skills and learn new procedures
Improved communication with design, construction, maintenance, planning, etc.	
Satisfaction providing best value for available funds	Training and education costs
Capability of assessing the implications of less funds, lower standards, etc. (as previously mentioned) but also capability of making the case for higher standards	Many activities are already being done but require some modification or retraining.

TABLE 18.2   Benefits of Infrastructure Management at the Agency Level

1. Project maintenance and rehabilitation needs and budgets for maintenance and rehabilitation
2. Justification for requesting funds
3. Measure effectiveness of dollars spent on the system
4. Control user cost by providing a known level of service
5. More effectiveness of dollars spent on the system
6. Improved methods of:
   - Planning
   - Design
   - Construction
   - Maintenance and rehabilitation
   - Research models
7. Improved safety
8. Priority programming for projects
9. Knowledge of statewide conditions and maintenance and rehabilitation needs for infrastructure
10. Improved serviceability level of infrastructure portfolios
11. Savings in user costs
12. Use of database to develop/improve performance models
13. Reduction of excessive rehabilitation costs because of timely action on maintenance and repairs

TABLE 18.3   Costs of Infrastructure Management at the Agency Level

1. Task force within the department to oversee and implement the IMS
2. Consulting services for developing the system based on the objectives and scope defined by IMS task force
3. Data Collection
   a. Agency personnel (engineers, technicians, equipment operators)
   b. Training
   c. Condition-monitoring equipment
      ■ vehicles for transportation
      ■ data loggers for condition surveys
      ■ automated distress survey devices
      ■ nondestructive testing equipment
      ■ roughness measuring equipment
      ■ drilling and coring equipment
      ■ construction acceptance (testing)
   d. Traffic-control services
   e. Travel and subsistence cost for data collection
   f. Traffic or use data (for example, AADT, ESAL, and WIM)
4. Data Processing
   a. Personnel (system analyst, programmers, and technicians)
   b. Equipment (computers, terminals, storage and backup devices, and calculators)
   c. Supplies
5. Data analysis and reports
   a. Personnel (engineers, system analyst, and technicians)
   b. Computers and terminals
   c. Supplies
6. System maintenance
   a. Personnel (engineers, system analyst, technicians, computer support staff)
   b. Equipment maintenance
7. Retraining of agency personnel associated with infrastructure management and IMS database
8. Administrative costs

TABLE 18.4   An Early Case Study: Economic Comparison of 1979 Priority List and a Subjective Programming Approach

Costs	Present (1979) dollars		
	IMS list	Needs study	Difference
Capital costs	$8,671,961	$8,627,988	$43,974
Maintenance costs	$1,804,406	$1,798,209	$6,197
Study costs*	$75,000	$0	$75,000
Total costs	$10,551,367	$10,426,153	$125,214
User benefits	$110,013,920	$98,847,056	$11,166,864
Net benefits	$99,462,553	$88,420,903	$11,041,650

*IMS cost based on approximate $200/mile. Note that the Needs Study has been assigned no study cost.

not been widely exploited within the U.S., although it is widely used in World Bank evaluations for developing countries.

Although some work has been done, there is still inadequate information about the benefits of timely maintenance and rehabilitation compared to its costs. Considerable work is needed in this area.

Another benefit of IMS is an improved awareness of the process of management. The authors have seen this occur during the teaching of pavement and infrastructure management graduate-level courses at the Universities of Texas, Waterloo, and Mississippi, where persons taking the course for a variety of reasons (some were assigned by their supervisor) turn into real advocates of IMS, and go back to their jobs with a renewed enthusiasm for providing good facilities for the public.

A final illustration of IMS benefits that are hard to quantify is the spin-off of IMS technology to other areas. After ten years of PMS use, for example, attention was turned to the development of BMS (bridge management systems), which has now become standard nationwide in the U.S. [Hudson 87].

18.5   Evaluation Methodologies

Some attention has been given to the various methodologies for comparing costs and benefits. The authors examined more than 20 references (cited at the end of this chapter) treating many of the most likely methods, ranging from discriminate analysis to general decision theories. Of the many generalized methodologies evaluated, the most likely candidates are reviewed here.

There is great conflict between goals and objectives in certain groups. No amount of rational data will eliminate all such conflict. It is impossible to please all individuals and groups playing their various roles at different times and places in the management process. Decision making must, however, be improved if better facilities are to be provided in a general context at the most economical cost.

18.5.1   Benefit-cost criterion

Perhaps the best-known criterion for measuring the efficiency of an activity is the benefit-cost analysis or the benefit-cost ratio. Efficiency is discussed in terms of benefit-cost analysis because other variations, such as the rate of return, are sufficiently similar to the benefit-cost analysis to have the same strengths and weaknesses. Because of its theoretical foundation, the benefit-cost analysis provides a conceptually sound basis for effective comparisons. In practice, however, many difficulties arise that tend to reduce its usefulness. The biggest drawback to benefit-cost analyses arises from the difficulty in breaking the factors into either the cost or benefit category, and the difficulty of actually measuring the true cost and the true benefit. There are many intangibles in benefit-cost analyses. These intangibles can be treated in three ways:

1. They may be rated subjectively and included in the analysis.
2. When subjective scaling is not possible, verbal descriptions of intangible benefits can be provided, in addition to the measured costs and benefits, and these used as aids by the decision-maker.
3. Unfortunately, the most common treatment for intangibles is to simply ignore them.

In other words, the preoccupation of the benefit-cost analysis with measurable impacts generally encourages the omission of intangible benefits. The result is a sort of inflexible narrowness that makes the benefit-cost approach incorrect and misleading at best.

Another problem associated with benefit-cost analyses are the issues of "who pays the cost" and "who receives the benefits." It is not always easy to ascertain whether the benefits and costs occur to the same stake-holders. For example, the improved programming of maintenance funds may be a benefit of IMS that accrues to the agency, and thus to the riding public or taxpayers. The cost, on the other hand, may involve a change in working assignments, or the requirement of additional training for various state personnel. They (the personnel) must be convinced to be willing to "pay this cost," even though the benefits accrue to someone else.

18.5.2   Excess benefits

There are many variations of cost-benefit analyses. One of these is the comparison of cost and benefits by calculating the excess of benefits over costs. A simple case-study comparison of this methodology is shown in Table 18.4. It has no particular value over other benefit-cost ratio methods and will not be considered further, except as a part of the benefit-cost-ratio concept.

18.5.3   Goal achievement

As one attempts to use the various methodologies for analyzing costs and benefits, and reviews criticisms of benefit-cost and similar evaluation procedures, one may be dismayed with the seemingly overwhelming complexity facing the decision-maker. One also gains some appreciation for the position of the politician or the manager (CEO) who must react to and give solutions for similar complex problems every day. The question arises: What, if anything, can be done to improve decision-making procedures? One technique for broadening the evaluation and decision-making process is known as "the goal-achievement method" [Dickey 82]. In this approach, potential alternatives are assessed in terms of impacts in comparison to the objectives proposed. Quantifiable measures are employed in this process, although some may be subjective and even probabilistic. In general, the procedure is to establish various criteria or "goals" for alternative methodologies. Quantitative measures or subjective estimates are then made for each criterion for each of the alternatives. These are then standardized and compared on the basis of a total score of 100 to see which alternatives most completely fulfill or achieve the goals of the decision-maker(s).

18.5.4   Cost-effectiveness technique

The cost-effectiveness technique (C/E) is an alternative approach to the goals-achievement procedure. C/E is a much less sophisticated procedure, and works on the basic premise that better decisions will arise when clear and relevant data are supplied to the decision-maker. No specific attempt is made to put all benefits and costs in common units, such as the dollar. In his book, Dickey [Dickey 82] quotes Thomas and Schofer on the cost-effectiveness approach:

> Because many of the consequences and outputs from a transportation system are intangible and otherwise difficult to evaluate in some common metric, the decisions regarding the conversion to a single dimension and hence the plan selection decisions are necessarily subjective in nature, at least at the present time.
>
> What might be more useful at this time is a technique for providing the kind of informational support for the selection among plans which recognizes the complex nature of these decisions. Such a decision-supporting framework does not attempt to *make* decisions, but instead *structures the information* required for making a subjective, but systematically *enlightened* choice. At the same time, however, the framework must be sufficiently flexible to permit the adoption of more sophisticated techniques, such as analytical methods for realistically implementing benefit-cost analysis or ranking schemes, when such techniques are appropriate.

Dickey lists three criteria that any framework for evaluation should satisfy:

1. It should be capable of assimilating benefit-cost and similar methodological results, in addition to other informational requirements.

2. It should have a strong orientation toward a system of values, goals, and objectives.

3. It should allow for the clear comparison of trade-offs or compromises between objectives, and/or should make explicit relative gains and losses from various alternatives.

Effectiveness is defined as "the degree to which an alternative achieves its objective." The definition, by itself, helps to overcome one of the major objections to the benefit-cost approach, in that goals are specified explicitly and are not covered by an all-encompassing "benefit" term. In benefit-cost analysis, for example, benefits are related to reductions in user-cost, user operation costs, and user time, but in a particular situation, any one of these factors may be of less concern. The objectives to be met may be associated with an entirely different set of factors specified explicitly, and not covered by an all-encompassing "benefit" term.

The value of the cost-effectiveness approach is threefold:

1. It simulates, to some extent, the process by which actual decisions are made.

2. It allows for clearer delegation of responsibility between analysts and decision-makers.

3. It makes it easier to provide relative information, structured in an understandable form so that the choice process is simplified.

18.5.5 Search and choice

In Manheim's work on transportation systems analysis [Manheim 69], he deals with "search and choice in transport systems analysis," and outlines what he calls PSP (Problem Solving Process) for use in dynamic modeling of decision-making. He outlines this process as shown in Figure 18.1.

Figure 18.1 Basic cycle of PSP (Problem Solving Process). [after Manheim 69].

According to Manheim, the focus of PSP is on actions. Because search and selection procedures concern the basic processes of generation and selections of actions, these procedures are at the heart of the PSP. However, there are a variety of other activities that occur in PSP to allow search and selection to operate, and to revise the context in which they operate. In particular, goal formulation and revision procedures are important.

Although the PSP activity is a very valid process for a decision-maker to go through, it does not appear at this time to be an appropriate methodology for use in the next phase of IMS.

18.5.6   Statistical decision theory

The world is a very uncertain place, but it is easy to forget this and become fascinated by quantitative data produced by complex models and elaborate calculations, such as benefit-cost analyses. In truth, there is always uncertainty in such analyses. Uncertainty in transportation is of at least three types: demand (such as traffic), technology, and goals. No matter how elaborate the prediction models are, or how much data are collected, there will always be uncertainty about predictions of future traffic and future pavement performance and life, as well as future maintenance costs and inflation. Statistical hierarchical decision processes are outlined by Manheim [Manheim 70]. He has developed a statistical decision approach to complicated transportation planning theories. However, from evaluation to date, it does not appear that this methodology is particularly applicable to IMS.

18.5.7   Discriminate analysis

Discriminate analysis and classification are multivariate techniques concerned with separating distinct sets of objects and with allocating new objects to previously defined groups. To obtain the best results, analysts seek discriminates whose numerical values are such that the collections are separated as much, or as distinctly, as possible. The goal of classification is to sort objects into two or more labeled classes. The emphasis is on deriving a rule or rules that can be used to optimally assign a new object to the labeled classes.

A function that separates may serve as an allocator, and conversely an allocatory rule may suggest a discriminatory procedure. In practice, the distinction between discrimination (or separation) and classification (or allocation) is not so clear. One of the objectives of conducting discriminate analysis would be to provide the basis for a classification rule.

The methodology of discriminate analysis, while useful in dealing with a large number of objects, does not appear to be appropriate for the study of IMS, and no further review is recommended [Lee 85].

18.5.8   Other methodologies

There are many other methodologies that have been used for decision making. Among these are terms such as benefits/risk analysis and "preference and value trade-offs." Within the confines of this book, a considerable amount of effort has been assigned to examining methodologies. No additional methodologies were found that appear to be adequate to the subject at hand.

Any evaluation used must be meaningful to the decision-maker. After all, the purpose is to present information that will be useful in convincing decision-makers to implement improved IMS methods. This concept of acceptability to the decision-maker must be given full attention in developing IMS.

18.5.9   Summary comments on methodologies

Three of the methodologies previously outlined above may be useful in clarifying the value of IMS. Certainly, benefit-cost analyses cannot be dismissed—comparing costs and benefits for a sample case-study network is an effective way of convincing decision-makers of the value of IMS. Cost-effectiveness techniques also bear additional study, and should be given strong consideration.

Finally, the general concept of goal-achievement methods should be explored. It is not clear exactly how the method might be applied, since it involves an examination of the goals of each decision-maker on an individual basis. This might be handled with hypothetical examples, if interviews with several decision-makers could be arranged for information-gathering purposes.

18.5.10   Data needs

It is valuable to address the data that would be needed for each of the methods recommended for study. In each of the three methods, costs data are needed: the costs of developing and using IMS, as well as the true costs of the components in the infrastructure site or network. Table 18.1 summarizes in four categories the types of cost data that would be included. Some of these data exist, but must be collected carefully. Other cost data will be easier to obtain, and more details on the kinds of cost information needed are given in Table 18.3.

In addition to cost data, which are common to all three methods, the other data needs vary among the methods as follows.

18.5.10.1   Costs-benefits.   For the costs-benefits method, data will be needed on the benefits outlined in Tables 18.1 and 18.2. It will be necessary to locate quantitative information on as many of the benefit classes as possible. In many cases it is difficult to locate exact financial information, and therefore estimates will have to suffice. It is also necessary to assess source and disposition of benefits; that is, "who pays the cost and who receives the benefits."

18.5.10.2 Goal achievement. The goal-achievement method will require inputs on the goals of CEOs in regard to their infrastructure network. The goals must be assessed in quantitative terms or in terms of criteria. They will take the form of factors such as "higher levels of serviceability," "longer average life," and "reduced user delay time." It is anticipated that several interviews with agency administrators will be needed to gather data on such goals and criteria.

18.5.10.3 Cost-effectiveness. Cost-effectiveness requires clearly structured cost data. More importantly, a set of goals must be specified explicitly; these are determined from decision-maker discussions and opinions. Final objectives must be set forth from the goals in clear, qualitative terms. Also, the rankings that CEOs assign among these objectives must be known.

18.6 Summary

It appears that adequate information can be gathered to make valid comparisons, which can be presented in a unique and meaningful way to the target audience.

Costs and benefits are hard to quantify. For this reason, cost-effectiveness and goal-achievement methods need to be examined in more detail.

Much is said in this chapter about better defining the benefits of IMS, and this should be done. It is clear, however, from the developments in this field over the past ten years that the overall benefits of good management are pervasive and large. This is clearly noted in the private sector, and must also be true in the public sector. The principles and methods provided in this book can help to fulfill this important goal.

References

[ASCE 82]   "Multiobjective Analysis in Water Resources," *Proceedings of the Engineering Foundation Conference,* ASCE Water Resources Planning and Management Division, New York, Nov. 1982.

[Dickey 82]   J. W. Dickey et al., *Metropolitan Transportation Planning,* 2nd ed., Hemisphere Publishing, McGraw-Hill Book Company, New York.

[Goicoechea 82]   A. Goicoechea, D. R. Hansen, and L. Duckstein, *Multiobjective Decision Analysis with Engineering and Business Applications,* John Wiley and Sons, New York, 1982.

[Haas 78]   R. Haas and W. R. Hudson, *Pavement Management Systems,* McGraw-Hill, New York, 1978.

[Haas 95]   R. Haas and W. R. Hudson, "The Customers for Pavement and Maximizing Their Satisfaction," *Proceedings, Canada Transportation Conference,* Victoria, B.C., Oct. 1995.

[Hudson 87]   S. W. Hudson, R. F. Carmichael III, L. O. Moser, and W. R. Hudson, "Bridge Management Systems," *National Cooperative Highway Research Program Report 300,* Transportation Research Board, National Research Council, Washington, D.C., Dec. 1987.

[Keeney 76]   R. L. Keeney and H. Raiffa, *Decisions with Multiple Objectives: Preferences and Value Tradeoffs,* John Wiley and Sons, New York, 1976.

[Lee 84]   H. Lee and W. R. Hudson, "Development of a Program Level Pavement Management System for Texas," *Research Report No. 307-3,* Center for Transportation Research, The University of Texas at Austin, Dec. 1984.

[Lee 85]   H. Lee, W. R. Hudson, and C. L. Saraf, "Development of Unified Ranking Systems for Rigid and Flexible Pavements in Texas," *Research Report No. 307-4F,* Center for Transportation Research, The University of Texas at Austin, Nov. 1985.

[Luce 57]   R. D. Luce and H. Raiffa, *Games and Decisions,* John Wiley and Sons, New York, 1957, Chapter 13.

[Manheim 69]   M. L. Manheim, "Search and Choice in Transport Systems Analysis," paper sponsored by Committee on Transportation System Evaluation and presented at the 48th Annual Transportation Research Board Meeting, *Transportation Research Record* 293, Washington, D.C., 1969, pp. 54–82.

[Manheim 70]   M. L. Manheim, "Decision Theories in Transportation Planning," *Special Report No. 108,* Highway Research Board, National Research Council, Washington, D.C., 1970.

[Manheim 75]   M. L. Manheim et al., "Transportation Decision-Making: A Guide to Social and Environmental Considerations," *National Cooperative Highway Research Program Report No. 156,* Transportation Research Board, National Research Council, Washington, D.C., 1975.

[Raiffa 68]   H. Raiffa, *Decision Analysis—Introductory Lectures on Choices under Uncertainty,* Addison-Wesley, Reading, Mass., 1968.

[Smeaton 81]   W. K. Smeaton, M. A. Karan, C. Bauman, and G. A. Thompson, *Case Illustration of Long-Term Pavement Management and Update,* Can. Tech. Asphalt Association, Nov. 1981.

[UCOWR 81]   "Risk-Benefit Analysis in Water Resources Planning and Management," *Proceedings, UCOWR-Sponsored Engineering Foundation Conference,* Y. Y. Haimes (ed.), Center for Large Scale Systems and Policy Analysis, Plenum Press, New York, 1981.

19

Future Directions

19.1 Introduction

No technical area, including infrastructure management, can remain static. IMS presents a kaleidoscope of opportunities for improvement and advancement due to changing technology, changing needs, and changing user demands. Numerous ideas could be presented here by the authors and then immediately expanded upon by the readers. Rather than belabor intricate detail, we have chosen to outline vision and future directions in four categories, to provide a semblance of order:

1. Advances in technology to assist the IMS process
2. Improvements in IMS resulting from its continued use (experience)
3. Advances made possible by the broader use of IMS methodology
4. Better education, implementation, and adoption of IMS

There are other ways of organizing these ideas, and there may be overlap in these four categories, but this will suffice to point out the new directions needed here.

19.2 Advances in Technology to Assist the IMS Process

Technology has advanced greatly in the 30 years since modern infrastructure management was initiated in the form of pavement management. Major developments include increased size and capability of computers and data-collection and storage activities. This change will continue, and during the next two decades the following specific technologies will have major impacts on IMS:

1. Geographical Information Systems (GIS) While GIS is already used in infrastructure management as illustrated herein, it will be greatly improved. It will have a dramatic impact in the coming decade as more user-friendly and generally applicable systems are developed and implemented.

2. Global Positioning Systems (GPS) Global Positioning Systems are already being used in a variety of civil-engineering infrastructure. The accuracy of these satellite-based systems is increasing by orders of magnitude annually while the cost is decreasing rapidly. The ability of infrastructure management staff to collect data in the field or perform maintenance and directly code it by location and identification through GPS will greatly enhance IMS.

3. Faster and More Convenient Databases Forty short years ago, punched cards provided the best access to data, along with mechanical sorting and high-speed linear magnetic tape (a useful medium, but one that provided no random access). Current random-access memory has increased its ability to store and process data by orders of magnitude. This improvement will continue as bubble memories and other new technologies are developed. A note of caution, however, is necessary, because it is tempting to invoke data pollution with such economical data-storage techniques. The collection, storage, and retention of vague or nonspecific data should be studiously avoided. One such example is the unnecessary storage of distress photographs. Photographs take hundreds of times more storage space than processed distress data, and distress photographs must be processed before they can be used. Thus, processing should take precedence over storage.

4. Smart Systems So-called smart systems (in this context) are those that have high-quality, long-lived sensors that can detect meaningful response in the facility and feed back that response data to adjust the characteristics of the system. Such systems have broad use in military and space, in guidance and aiming devices. They can potentially be applied in infrastructure management in certain types of maintenance feedback activities. Many other uses will also be developed as they become available with more reliability at a lower cost.

5. Better Communications Technology Communications is a critical part of infrastructure management. Communications technology will continue to improve and find applications in IMS. Again, however, it will be critical to use the technology wisely. We all can cite examples of misuse of e-mail and the Internet.

19.3 Improvements in IMS Resulting from Its Continued Use

When an IMS is used by a specific agency, improvements will accrue within that agency itself, just as the ability of an athlete or musician improves with practice and experience.

1. Performance Models At its inception, any IMS starts with approximate performance models. As the agency gains five, ten, or more years of experience and history in the database for the segments or units within the management system, performance-history data becomes available for developing improved performance models. Thus, the results of the IMS process improve with time.

2. Better Segmentation Within many IMSs, it is necessary to subdivide the network into sections and subsections. This segmentation is often approximate at the beginning of the process, but improves with time and use of the system, including dynamic segmentation.

3. Identification—Location Since most networks or portfolios of infrastructure are pre-existing at the beginning of an IMS, it is not always possible to identify these separate parts correctly and completely. After the IMS has been used for several years, the identification and location information improves within the system.

4. Add-on Subsystems As an IMS is used continuously, it is possible to add additional subsystems and therefore greatly improve the overall quality of the IMS.

5. Flexible Utility—User-friendly Output reports and utility of IMS can be greatly improved over years of use, particularly within an existing agency. It is not always possible to know precisely the user needs at the birth of the system, but clarity comes with age.

6. Improved Quantification of Qualitative Factors A limitation on the use of any IMS is the ability to account for qualitative factors, such as judgment and public opinion. These factors are critical to the IMS process, and the ability to quantify knowledge of the proper relationships is greatly improved by continuous use of the system.

19.4 Advances Made Possible by the Broader Use of IMS Methodology

As IMS methodology becomes more widely used in diverse areas, such as water and waste water, it becomes possible to integrate new technologies and ideas into the process. This is an important way in which some of the following new directions will be developed.

1. Evaluation and Use of New Materials New materials can be integrated into an IMS in a rational way; however, if no IMS exists, there is often no way to compare the relative value of the new material to existing, less costly materials.

2. New Technologies The same is true for new technologies. An IMS framework provides an ideal way to apply new technologies that have little chance of proving their worth under conventional design and maintenance operations, except by expensive trial and error.

3. Incentive to Develop New Materials Experience has shown that once materials and technology experts know that IMS exists, they will speed up the development of new technologies and new materials because the developers or inventors have a better chance of seeing their development or invention used effectively.

4. Institutional Improvements The fact that IMS is used effectively in other agencies will encourage institutional issue improvements among other new potential users. That is, agencies that are not now using an integrated systems approach may be able to see the benefit of such an approach among their peers in other similar agencies.

5. Integration of Additional Facilities An agency, such as an urban center or city, may start with a simple IMS involving its streets and bridges. However, use of that system can encourage the expansion and addition of water facilities, waste water, sewer, and associated utilities in the right-of-way, such as phone, electric, and gas.

6. Application of IMS to Unitized Facilities Currently, not all airports, park agencies, and airport centers or port facilities use the benefit of an integrated management system for their infrastructure. The availability of such systems in practical use among their peers and on similar facilities in the future will encourage the adoption and use of such facilities by previous nonusers. This effectively ties in with the geographically central location of the unitized facility.

7. Automated Data Collection As broader use of IMS develops, it will become easier to obtain automated data-collection services from service-providers and contractors. When the market becomes large enough, this service can be economically offered by sharing among several user groups, and the price will drop.

8. Improved Resource Allocation between Existing and New Facilities As IMS is more broadly used, it will become an effective tool for an agency to allocate its resources between maintaining existing infrastructure and building new infrastructure. This is most often an ad hoc activity at the present time, with political forces often pushing for new facilities that offer strong political impact in lieu of much-needed, but less glamorous, maintenance funding.

9. Improved Resource Allocation among Types of Infrastructure Improved IMS also will make it possible to better allocate resources among various types of infrastructure: roads versus sewers versus water, and so forth. Currently, such allocations are almost always politically motivated.

10. Improved Consideration of Benefits   Broader use of IMS, combined with the desire to evaluate among and across infrastructure types, will encourage better consideration of user benefits as part of the resource-allocation equation. This will allow agencies to move closer and closer to optimal use of their funds, and offer an improved quality of life for the taxpayers, who are the true owners of the infrastructure.

19.5   Better Education, Implementation, and Adaptation of IMS Technology

IMS technology will lag in its use until potential users become fully educated about its benefits. Current college graduates are told little or nothing about the need to integrate design, maintenance, and rehabilitation. They are taught design and perhaps construction quality, but little else about the total life-cycle of facilities. Education in the form of short courses, college-level courses, and graduate work in IMS will encourage improvements in and wider use of IMS by engineers and managers. In this way, educational workshops and short courses will encourage the exchange of information among peer groups and will foster the spread of IMS from agency to agency.

At the administrative level, courses of one- or two-day duration to apprise elected officials and administrators of the benefits of IMS as a decision-making tool also will broaden its acceptability and adoption. Of particular value is the introduction of an undergraduate course in IMS in civil-engineering curricula. Such a course will introduce young civil engineers to the need for improved maintenance and rehabilitation engineering. Currently, the emphasis is very strongly on design. At least 90 percent of all civil-engineering graduates now deal with maintenance and rehabilitation in their job in some fashion.

19.6   Summary

Ultimately, the future of IMS lies with its users and its potential users, like the readers of this book. Anyone who has read this book, or a large portion of it and has now reached this section, is clearly a potential user of IMS technology. Hopefully, readers will apply this technology and make it known to their peers and other potential users. There are many sources of additional IMS information in the literature and at current conferences and meetings. This information may come in bits and pieces, but readers should take the opportunity to use such information as they continue their quest to improve the quality of infrastructure management of public facilities, both their own agencies and throughout the world.

Index

Index

ABOUT THE AUTHORS

W. RONALD HUDSON is the Dewitt C. Greer Centennial
Professor of Civil Engineering at the University of Texas at
Austin. A pioneer in developing and applying management
systems to the infrastructure problems of bridges and high-
ways, Dr. Hudson is co-author of *Modern Pavement
Management*.

RALPH HAAS is the Norman McLeod Engineering Professor
and Distinguished Professor Emeritus at the University of
Waterloo in Ontario, Canada. An innovator in pavement
design and management, Dr. Haas is co-author of *Modern
Pavement Management*.

WAHEED UDDIN is Assistant Professor of Civil Engineering at
the University of Mississipi. Dr. Uddin has served as a pavement expert for the UN, and has contributed to highway
management projects in the U.S. and many other countries.